VOLUME 30

BOEING C-17A
GLOBEMASTER III

By Bill Norton

specialtypress
PUBLISHERS AND WHOLESALERS

COPYRIGHT © 2001 BILL NORTON

Published by
Specialty Press Publishers and Wholesalers
11605 Kost Dam Road
North Branch, MN 55056
United States of America
(651) 583-3239

Distributed in the UK and Europe by
Airlife Publishing Ltd.
101 Longden Road
Shrewsbury
SY3 9EB
England

ISBN 1-58007-040-X

All rights reserved. No part of this book may be reproduced or transmitted in any form or by any means, electronic or mechanical including photocopying, recording, or by any information storage and retrieval system, without permission from the Publisher in writing.

Material contained in this book is intended for historical and entertainment value only, and is not to be construed as usable for aircraft or component restoration, maintenance, or use.

Printed in China

Front Cover: *The many interior and exterior lights, including those in areas serviced by mechanics, are clear in this sunset image of P-35. Nicely outlined are the forward nose gear doors that open during gear transit. The subdued seal markings aft of the crew door are those of Air Mobility Command and the Air Force Reserve. (Boeing)*

Back Cover (Left Top): *A 437th Airlift Wing Globemaster III, P-7, is seen over the South Carolina coast near its home base of Charleston AFB. Its low altitude is evident by the shadow cast on the surf below. The outfit's attractive tail logo was frequently seen during the C-17's first years while it was the only operator of the big airlifter. (DoD)*

Back Cover (Right Top): *Some of the external features of the Pratt & Whitney F117-PW-100 engine are depicted here, including the Electronic Engine Control (EEC), High Pressure Compressor (HPC) case, Low Pressure Turbine (LPT), and High Pressure Turbine (HPT) cases. (Boeing)*

Back Cover (Lower): *Equipment pallets depart the ramp of a very dirty P-12 in the new dual row configuration. The drogue (off the top of the image) is pulling out the extraction parachute on a pallet that has flipped inverted. Note that the aircraft's slats and flaps are extended for this airdrop condition. (DoD)*

Title Page: *Three of AFFTC's flight test C-17s (from front to rear P-1, P-2, and P-3) taxi together at Edwards AFB. The article hanging from the tail of P-1 is a flight test trailing cone for measuring static air pressures in flight. The tubing is reeled out to about 150 feet aft of the aircraft in flight. (McDonnell Douglas)*

TABLE OF CONTENTS

BOEING C-17A GLOBEMASTER III

PREFACE

The C-17A Globemaster III is still a young aircraft, being less than 10 years since its first flight and with production less than halfway through its presently funded run. However, its gestation is a fascinating study in the development and acquisition of U.S. military aircraft. Combining the strategic and tactical airlift missions in one airframe with Short Takeoff and Landing (STOL) capability augmented with a fly-by-wire flight control system makes the C-17 an especially interesting transport aircraft. Its mission is varied and dynamic and its short service record to date has been one suggesting a long and noteworthy future. My own experience with the C-17 dates back to two years before first flight and through the flight test program where I served as a lead U.S. Air Force flight test engineer and USAF officer. I think readers will agree that this is an auspicious time to detail the Globemaster III.

The C-17 program initially resided with the Douglas Aircraft Company, McDonnell Douglas Corporation, before the C-17 team became part of the newly formed McDonnell Douglas Aerospace (MDA), Transport Aircraft Division, in 1992. The C-17 was well established in production when McDonnell Douglas and Boeing merged on 1 August 1997, with the latter name carried on. For the sake of historical accuracy the appropriate name is used in the relevant portions of the text.

Many individuals, too numerous to name here, must be thanked for contributions that made this book possible. Many individuals in the System Program Office and U.S. Air Force Public Affairs offices proved most helpful with information and photos. Len Tavernetti at Boeing lent invaluable support to the project, especially in illuminating the historical perspective and checking facts. The personnel of the C-17 flight test team provided generously of their time and knowledge. Lt. Col. John Norton, a C-17 pilot and the author's brother, spent many hours reviewing the manuscript and explaining the complexities of military airlift. A hearty thanks is extended to all for their help.

This book is dedicated to all of the contractor and government personnel who gave so much of themselves over many years to make the C-17A a reality. Their perseverance and conviction through many crises, large and small, has been a constant inspiration.

Bill Norton
2000

The author (right) and his brother, C-17 pilot Lt. Col. John Norton (left) both contributed to making the C-17 a reality. The author served as a lead flight test engineer on the project while John served as an operational test pilot and later a staff officer working tanker/airlift issues at the Pentagon. John was also an essential contributor to this book. (Bill Norton)

INCEPTION

The Lockheed C-5 Galaxy entered U.S. Air Force service in 1973 with only 81 copies produced. Prior to this it was 1964 when the Lockheed C-141 Starlifter was introduced and 1956 for the C-130 Hercules. Clearly the introduction of new airlifter aircraft is more rare than fighter types as the mission seldom changes dramatically to demand an advanced design. However, by the mid-1970s changes in the military posture of the United States, anti-aircraft threats driving aircraft to fly at low altitude where buffet loads are more severe, aerospace advances that greatly aid complex missions, and a requirement for reduced operating costs provided the impetus for another U.S. Air Force transport aircraft. Still, it would be 1985 before a contract was issued to begin full-scale development of the advanced airlifter, the C-17A Globemaster III.

THE AMST PROGRAM

The C-17 had its beginning in the Advanced Medium STOL (Short Takeoff and Landing) Transport, or AMST, program. The AMST grew out of a 1968 requirement issued by the U.S. Air Force for a tactical transport capable of delivering wide U.S. Army vehicles to forward-deployed troops and a study by the U.S. Air Force investigating the potential of STOL transports. These efforts prompted a technology demonstration when, in 1972, the U.S. Air Force issued a request for proposal (RFP) for a STOL aircraft in the class of the venerable C-130 but with significant improvements in performance. Boeing and McDonnell Douglas were chosen to each prepare two AMST prototypes, YC-14s and YC-15s respectively, for a fly-off competition.

Douglas Aircraft Company (DAC) was to design and build the YC-15. Douglas' history with U.S. military transports began in 1925 with the C-1, the first formal cargo/transport aircraft bought by the U.S. Army. The most famous Douglas product was the C-47, the militarized DC-3. Over 10,000 were built during World War II during which the C-47 served with amazing utility. Immediately following the war DAC brought forth the little-known C-74 Globemaster cargo transport. Only 14 of the C-74s were built, but the design served as the basis for the double-deck C-124 Globemaster II. There were 448 of this versatile aircraft that served the U.S. Air Force during the Korean and Vietnam wars. The last effort was the giant Douglas C-133 Cargomaster that carried on the heritage in the late '50s as a turbo-prop transport specially designed to transport large rockets.

The requirements for the AMST aircraft were as demanding as those the C-17 would eventually be called on to meet. The proposal mandated a wide-body cargo compartment floor area of 6,125 square feet — providing 67 percent more cargo capacity than the C-130. The STOL aircraft was to routinely operate into and out of a 2,000-foot landing zone (LZ) with a 27,000-pound payload and at least half of its internal fuel. An alternate load of 53,000 pounds and unimproved field operations were

The C-17A Globemaster III is the largest STOL aircraft ever built and fielded. It is also the first new U.S. Air Force airlifter in nearly two decades and the result of an acquisition process set in motion even earlier. (Boeing)

The C-74 Globemaster was a wartime development of the Douglas DC-4, although it did not fly until after the end of hostilities. Although only a dozen entered service, they were the largest land-based transport in the world at the time. It is seen here in its "bug-eyed" configuration with two separate canopies for the two pilots. (Boeing)

The C-124 Globemaster II was a further development of the C-74. The deep fuselage accommodated large loads brought aboard through clamshell doors under the nose. With 448 copies and service into the 1970s, "Ol' Shakey" was a worthy predecessor to the C-17. (Boeing)

The impressive C-133 Cargomaster was the last purpose-built airlifter McDonnell Douglas manufactured for the U.S. military before taking on the C-17 program. In the 1980s, McDonnell Douglas produced the KC-10A tanker as a conversion of the DC-10 airliner. (Boeing)

WARBIRD**TECH**
S E R I E S

requested. The aircraft was to possess a 400 nautical mile radius of action with a 2,600 nautical mile ferry range.

The STOL performance requirements compelled the AMST designers to apply powered (or propulsive) lift concepts, a design approach that remains rare even today. This involves turning the engine exhaust flow downward to produce a vertical thrust vector in addition to the usual wing lift. By using the wing's trailing edge flaps to affect the flow turning, wing lift is enhanced and no other heavy mechanisms, such as rotating exhaust nozzles, are necessary.

One consequence of powered lift is that a change in the manner of flight path control is needed during approach to landing. This is called the "backside" technique — referring to the low-speed side of the velocity-versus-drag or the power-required curve — in which thrust controls flight path angle and pitch attitude controls airspeed. This is the reverse of more conventional aircraft and requires additional training to master. The principal advantages of flying on the backside are lower landing speeds, steeper approaches, and improved touchdown accuracy; altogether reducing landing distance. The backside approach allows descent angles up to 6 degrees, versus the standard 2.5 to 3 degrees, to help facilitate a short landing.

BOEING YC-14

The Boeing team adopted the Upper Surface Blowing (USB) powered lift method. In this design the engines exhaust over the top of the wing. With the flaps positioned down, the airflow stays attached to the flap's upper surface and turns to follow its contour as a result of the Coanda Effect. Boeing used two 51,000-pound thrust General Electric F103-GE-100 turbofans for their design.

The 121-foot long YC-14 had a 129-foot span straight wing with 6 flap segments. The maximum takeoff gross weight was 225,000 pounds. The engines exhausted over the inboard of three flap segments per wing through a D-nozzle that mixed the fan and core exhausts. The titanium heat shield decking aft of the exhaust included four large vortex generators (VGs) to improve flow turning. These VGs normally laid flat but rotated to the vertical when the flaps extended beyond 36

The origins of the C-17 lay in the AMST program that included the Boeing YC-14 as one of its competitors. This transport used the Upper Surface Blowing (USB) technique to produce STOL performance. The engine exhaust ducts and reversers are shown ahead of the extended flaps in this image of a YC-14 under construction. (USAF)

The two prototype YC-14s completed a flight test program that demonstrated the feasibility of a wide-body STOL tactical transport, but the type was not taken up for production. The USB approach also saw later application in the Antonov An-72 and a number of modified testbed aircraft including a Kawasaki C-1 and NASA Dash 8 QSRA. (AFFTC)

degrees. The flaps could be deflected to 70 degrees and managed to turn the engine efflux through 60 degrees. Forward of the nozzle was a clamshell thrust reverser that directed the reversed exhaust up and forward but could not be used while airborne. Flight control was via a mechanical system with a three-channel digital electronic flight control system (EFCS) for stability augmentation and hands-off attitude hold. It automatically modulated the blown flap segments independent of the other flaps for the desired STOL performance.

To assist in precision STOL approaches, a heads-up display (HUD), then a novel installation for a transport aircraft, was installed with a Visual Approach Monitor (VAM) feature displaying the flight path vector. The Electronic Attitude Director Indicator (EADI) was another such aid. This consisted of a television camera mounted in the lower nose of the aircraft producing a view of the landing area on a screen in the instrument panel. The aircraft's attitude relative to the touchdown zone was superimposed on this image as well as the aircraft's flight path vector. An autobraking system provided braking immediately upon landing without pilot action, as well as deploying the spoilers.

The first Boeing YC-14 (72-01873) made its maiden flight from Boeing Field, Seattle, on 9 August 1976, followed by its companion (72-01874) on 21 October. Both were eventually ferried to the U.S. Air Force Flight Test Center (AFFTC), Edwards AFB, California, for a fast-paced but extensive service evaluation which included STOL operations on semi-prepared fields (unpaved but moderately dense and smooth surfaces), cargo operations, and airdrops.

Pilots found the YC-14 an exceedingly pleasant aircraft to fly and an exceptionally stable aerial delivery platform. The USB, the braking systems, and the considerable excess thrust were instrumental in producing outstanding STOL performance. The aircraft easily met the 2,000-foot landing goal, even with 53,000 pounds of payload, and demonstrated takeoff distances averaging 1,500 feet. Steep approach flight path angles of up to six degrees were routinely flown. The VAM was useful, particularly at night or in adverse weather conditions, but was not con-

sidered essential. The EADI was found too limited in resolution. The pilots also concluded that the control wheel, necessary in a fully mechanical control system or one without boosted controls, should be substituted with a stick for an aircraft like the YC-14 with a fully powered and augmented control system. In fact, during crosswind operations the wheel was awkward, conflicting with the pilots' knees and obstructing a portion of the instrument panel.

On the negative side, the reversers were deemed too slow in deployment. Even shorter landing rollout distances were possible if this were corrected. The aircraft was pitch sensitive since the engines were mounted so high above the vertical center of gravity (CG). Excessive drag from the exhaust flow over the blown flap was endemic to the design, contributing to the failure in meeting the cruise performance goal.

McDONNELL DOUGLAS YC-15

McDonnell Douglas adopted the Externally Blown Flaps (EBF) concept for the competing design, involving the interaction of engine efflux with slotted flaps. The positioning of the engine nacelles directly in line with the lower surface of the deployed flaps has the advantage of reducing cruise drag and simplifying maintenance access. However, this method requires flaps capable of withstanding the temperature and acoustic loading of the jet impingement. The YC-15 became the first jet-powered aircraft to utilize the EBF design.

Douglas chose to use four 16,000-pound thrust Pratt & Whitney JT8D-17 turbofans. Engineers estimated that the re-directed exhaust provided 20 percent of the lift during STOL operations, the augmented flap lift provided another 26 percent, and

the basic wing/flap combination contributed the final 54 percent. The YC-15 also employed direct lift control (DLC) by which the pilot could quickly alter the lift produced by the wing, for rapidly changing the flight path angle, by adjusting wing spoiler position. This initially involved modulating spoiler position.

The YC-15 measured 124 feet in length with a 110-foot straight wing. With an empty weight of about 115,000 pounds, the YC-15 had a maximum gross weight of 216,000 pounds. The two-segment flaps were made mostly of titanium with a forward vane and aft major panel separated by a slot. The spoiler panels ahead of the flaps provided another trailing edge slot when stowed with the flaps extended. The mechanical flight control system featured a dual-channel Stability and Control Augmentation System (SCAS) that enhanced the natural

McDonnell Douglas was the other contractor in the AMST program, offering its YC-15 STOL aircraft. The first example (72-01875) is shown climbing out on its maiden flight from the Douglas Aircraft Company plant in Long Beach, California, on 26 August 1975. It is a direct ancestor of the C-17. (USAF)

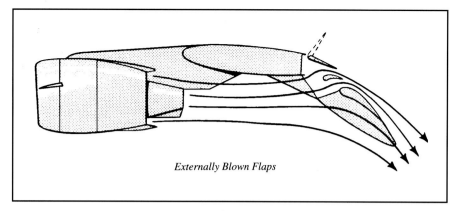

Externally Blown Flaps

The YC-15 used the Externally Blown Flaps (EBF) principle to produce its STOL performance. As illustrated here with the C-17 configuration, engine efflux impinges on the slotted flaps directly aft of the exhaust to increase overall lift and turn the flow downward. (Boeing)

stability of the aircraft as well as provided hands-off attitude hold and autopilot functions. Spoilers were biased up during STOL approach and driven up or down to provide rapid airspeed and altitude changes as the drag increment was added or subtracted.

An automatic ground spoiler (AGS) feature deployed the spoilers on landing to reduce rebound tendencies and to improve short field capability. Thrust reversers directed engine exhaust up and forward for unimpaired ground operations around the aircraft with engines operating. The reversers enabled aircraft backing on the ground and could be used in

The YC-15's ability to airlift wide tactical vehicles is demonstrated in this ground test. Steep approaches and operations on unimproved fields were also explored by the AMST competitors. (USAF)

flight for rapid deceleration. McDonnell Douglas also experimented with a pilot HUD on the YC-15 to improve touchdown accuracy.

The initial McDonnell Douglas YC-15 (72-01875) was flown on 26 August 1975 from the Douglas plant in Long Beach, California, to Edwards AFB. The second example (72-01876) followed on 5 December. Considering the new technologies it incorporated, the haste with which it was built, and the limitations imposed on an unproved design, the YC-15 performed remarkably well

ment during high temperature conditions. Delaying flaps deployment until late in the takeoff roll was experimented with and succeeded in improving takeoff performance. McDonnell Douglas struggled with repeated cracking in the blown flap material. Range was also deficient due to higher than anticipated drag.

Valuable lessons about STOL transport operations and the backside approach method were obtained from the YC-15. The flight and ground thrust reversing capability was found beneficial for the tactical

in a short period of time plus gleaning data and experience directly applicable to future STOL aircraft. However, the competition was terminated in January 1978 and the demonstration program wrapped up at the end of that year. The U.S. Air Force budget simply would not allow for purchase of a new Military Airlift Command (MAC) transport aircraft at that time. However, the requirement to expand the airlift fleet's capabilities was unquestioned. On the same day that the AMST was terminated a new program was initiated.

Although the AMST was principally a technology demonstration program, there were hopes a full-scale development contract would result. The defense budget would not support such an acquisition at the time and it would be more than 10 years before a widebody STOL transport, the C-17A, would take to the air. The original YC-15, like the other AMST prototypes, came to rest in the Pima Air Museum and Davis Monthan AFB; all sans their engines. (Bill Norton)

during its test program. The machine handled pleasantly in the air, exhibiting only minor flying quality deficiencies, and was easily maneuvered on the ground. Cargo loading, aerial delivery, low altitude parachute extraction system (LAPES), paratrooper drops, in-flight refueling, and semiprepared airfield operations were all demonstrated.

The YC-15 also suffered some inadequacies. Single-engine takeoff and landing performance was unacceptable and the aircraft was unable to meet its 2,000-foot landing require-

mission. The HUD as a STOL approach aid was considered mandatory, particularly during steep approaches. The DLC spoiler bias was reduced and eventually eliminated because of the unacceptable drag penalty. The AGS proved extremely effective and, when combined with the outstanding braking system, greatly facilitated short field landings. The YC-15 pilots also stated their preference for a stick rather than a yoke.

Both AMST prototypes performed well, producing promising designs

THE C-X PROGRAM

The need for a new U.S. Air Force transport became more urgent when the Rapid Deployment Force mission was formulated in the late '70s. The aircraft would be instrumental in allowing MAC to close what was clearly an airlift shortfall in plans to support a major war in Europe or the Middle East. After the end of the Cold War this evolved into a requirement to support two nearly simultaneous major regional conflicts. Also, with the contraction of U.S. presence overseas, airlift would play an even more important role in future conflicts.

The requirements for the new airlifter bore many similarities to the AMST, but the U.S. Air Force now sought a high capacity aircraft to augment the C-5 fleet. Like the Galaxy, it would accommodate both oversized (essentially wider than a normal cargo pallet) and outsized (essentially non-palletized cargo such as rolling stock and helicopters) cargo. The whole focus of the specification was economy through efficiency. The new aircraft would fly over strategic distances — the intertheater mission — that only the C-5 and C-141 could meet. However, it would also fill the short-distance intratheater tactical mission of the C-141 and C-130 with airdrop and STOL operations on austere (little or no facilities) semiprepared airfields.

Furthermore, the new aircraft's payload could be flown to forward deployment areas from the continental United States (CONUS) — termed "direct delivery" — rather than just to major airports from which the cargo must then be transshipped via C-130, frequently after break down and reconfiguration. Projections were that a major European war would require at least 66 million ton-miles of intertheater airlift per day and 16,000 tons per day intratheater, with the C-17 contributing to both. The aircraft would carry four times the payload of a C-130 yet land at fields only the Hercules could get into. Although envisioned as an aircraft between the C-141 and C-5 in size, it was to have exceptional maneuverability on the ground to permit loading and unloading operations on confined aprons with more of the aircraft in the available area.

The strategic and tactical delivery missions had never before been successfully integrated, much less in the largest STOL aircraft ever conceived, and these aspects constituted risk for the program dubbed C-X (Cargo - Experimental). However, the combined mission requirements also promised an airlifter of tremendous operational flexibility allowing MAC's airlift mission to be performed with fewer resources.

The U.S. Air Force did not dictate a particular design but rather specified the mission needs and let the design teams devise their best solutions. They requested a "rugged, reliable workhorse that is simple to maintain and operate. . . . Undue complexity or technical risk will be regarded as poor design." This and the historical advice against developing an engine and airframe simultaneously led the U.S. Air Force to require that "engines for the C-X system will be FAA certified prior to first flight." They also stated that "of primary importance is the effectiveness with which the system can perform the prescribed operational mission — considering performance, availability, reliability, crew compartment utility, system supportability, and safety. . . . A key consideration will be the Government's estimate of the most probable cost to develop, acquire, and support the operation of the C-X aircraft over a 20-year period."

The C-X RFP went out on 15 October 1980 and four proposals were submitted on the first day of 1981. The two Lockheed proposals included a modified Galaxy, designated the C-5M and the other looked much like an expanded C-141 with externally blown triple-slotted flaps. Boeing's proposal was an enlarged YC-14 design with a third engine added above the aft fuselage. The McDonnell Douglas aircraft was a growth of the YC-15 design featuring a swept wing with winglets and four large turbofan engines.

The successful McDonnell Douglas entry into the U.S. Air Force's C-X competition was essentially an enlarged YC-15, again employing EBF. This concept painting of what would become the C-17 shows that the design changed little in appearance from concept to hardware. The starboard landing gear sponson was later extended forward to house the APU, originally intended to be in the tail cone. (DoD)

On 28 August 1981 the McDonnell Douglas proposal, then called Model D-9000, was pronounced the winner and designated the C-17. Because of the AMST experience, the development of the new aircraft, using existing technology, was considered low risk and the program would be accelerated. First flight was planned for July 1985 and Initial Operational Capability (IOC) for September 1986. In early 1993 the U.S. Air Force Air Mobility Command (AMC) announced that the C-17 would be known as the Globemaster III. (MAC was disbanded on 1 June 1992 with its elements becoming part of AMC.)

Immediately after the C-17 competition, proposals for addressing the airlift capacity shortfall more expeditiously were approved. These proposals eventually took form with the ongoing C-141 stretch (C-141B) program, the purchase of 50 new C-5Bs, the continued production of the McDonnell Douglas KC-10 Extender tanker/transport, and the strengthening of the Civil Reserve Air Fleet. There was also a move afoot to equip an airlift wing with used Boeing 747F freighters, tentatively dubbed the C-19, that came to nothing. These measures provided an additional 20 million ton-miles of capacity but still left a 25 percent airlift shortfall. The requirement for the C-17 remained sound but the budget was again too constrained to support a new development effort and Congress directed that the program move ahead slowly. The consequence was a four-year stretch-out of the effort.

Between the selection of the prime contractor and the award of the Full Scale Engineering Development (FSED) contract, smaller engineering efforts were funded. This included a

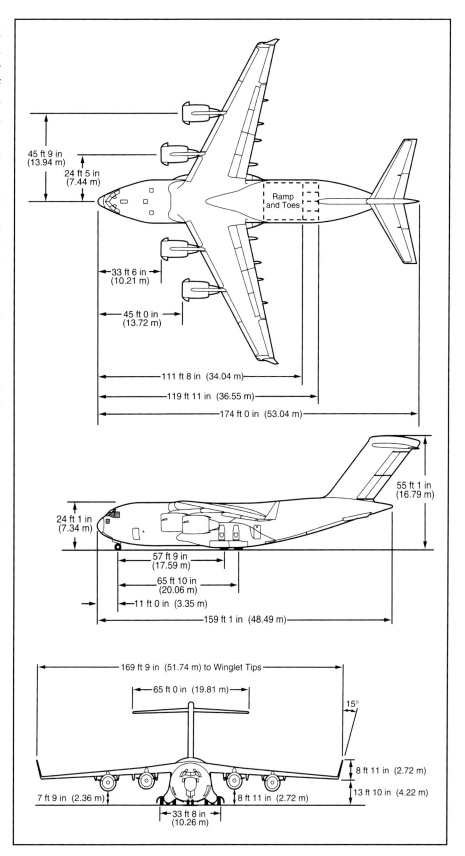

The C-17 is a large aircraft, unusually so for the tactical airlift tasks it performs. Although appearing "stubby" in some aspects, its dimensions are ideal for maneuvering within confined apron space. (Boeing)

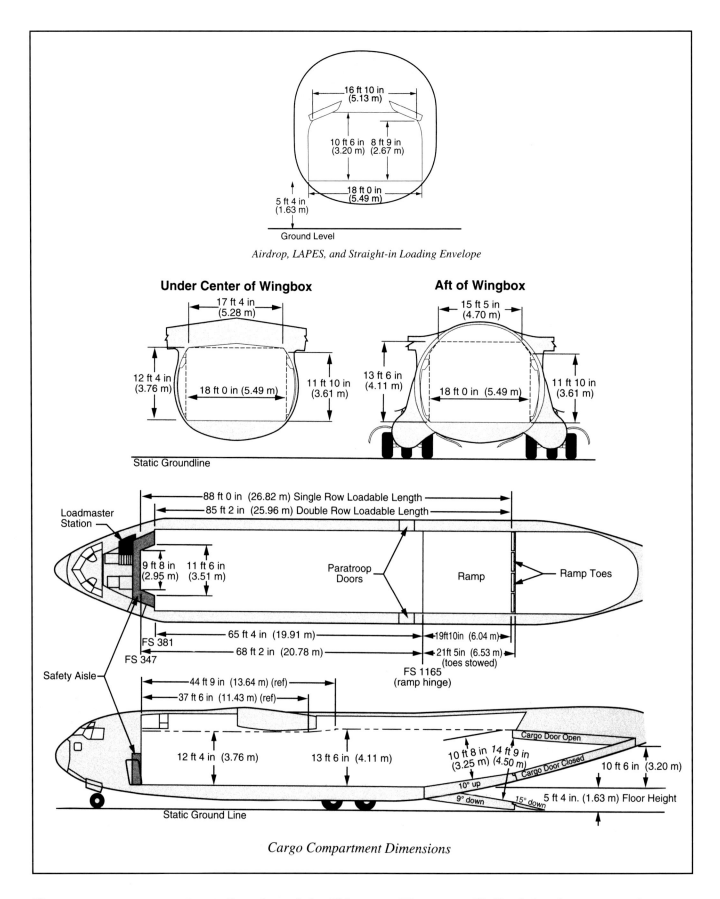

Airdrop, LAPES, and Straight-in Loading Envelope

Cargo Compartment Dimensions

The generous cargo compartment dimensions of the Globemaster III were specifically designed to accommodate many configurations of palletized loads, U.S. Army vehicles, and helicopters. (Boeing)

$31.6 million initial low-level development contract, promulgated on 23 July 1982, to allow early design studies and laboratory work to proceed. Among this labor was the commercial certification program for the selected powerplant and a revision of the 38,400-pound thrust Pratt & Whitney PW2037, which was already planned for operation on the Boeing 757 civilian airliner. The effort began as early as December 1979 and a suitability ground demonstration for the C-17 application was performed in 1983. The FAA certification of the engine was received on 28 December of that year. With the C-17 design maturing, further revisions were introduced and testing resumed in October 1985. The powerplant was certified again in 1987 as the PW2040, rated at 41,700 pounds of thrust, with the military designation F117-PW-100.

The $3.387 billion C-17 FSED contract was finally let on the last day of 1985 with a first flight planned for March 1990 and an IOC in April 1992 after sixteen aircraft had been delivered. McDonnell Douglas was to develop and build a single flight test article and required ground test airframes. Subsequent contract modifications funded initial production aircraft.

The C-17 was initially intended to complement existing C-130s, C-141s, and C-5s, many of which would be more than a quarter century old when the new jet was introduced. However, in the late 1980s the C-141 fleet began to experience wing fatigue cracks and other failures, partially accelerated by performing missions intended for the delayed C-17. This mandated an expensive retrofit of new centerwing boxes for many of the machines and retirement for others. The intensive Oper-

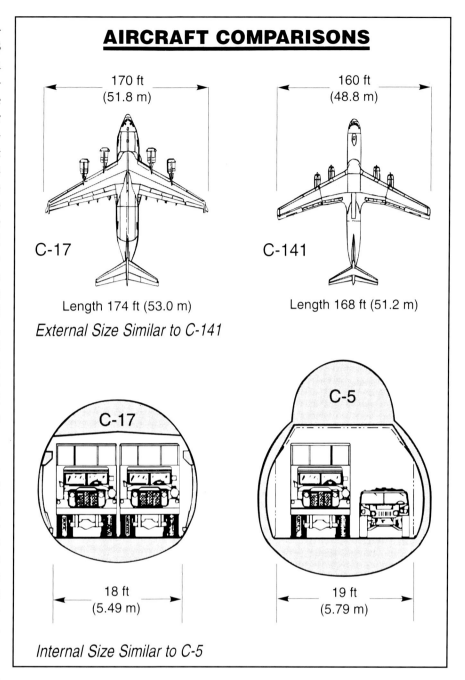

AIRCRAFT COMPARISONS

170 ft (51.8 m)

C-17

Length 174 ft (53.0 m)

External Size Similar to C-141

160 ft (48.8 m)

C-141

Length 168 ft (51.2 m)

C-17

18 ft (5.49 m)

C-5

19 ft (5.79 m)

Internal Size Similar to C-5

Although about as long as the C-141B, the C-17's wide-body cargo compartment approaches that of the C-5, with even some advantages on the Galaxy. However, the overall capacity of the strategic C-5 remains far ahead of its contemporaries. (Boeing)

ation Desert Shield airlift cut deeply into the remaining airframe life of the C-141s. Consequently, when the C-17 was finally introduced into the U.S. Air Force fleet the then-anticipated 120 Globemaster IIIs were to replace 234 C-141Bs and many older C-130s.

The slow start of the C-17 program and the surging aerospace market played havoc with McDonnell Douglas retaining sufficient talented engineers. The contractor was also encountering its share of challenges with the design including the change from a mechanical to a fly-by-wire flight control system, a change in the flight control software vendor, and a major weight reduction program to eliminate 20,000 pounds from the notional aircraft. The program was "re-baselined" in November 1989 to reflect budgetary decisions and significant program slips brought on by the engineering development delays.

The C-17 program's cost and need were repeatedly questioned during national budget debates. However, with a deficient airlifter fleet and promised benefits of the C-17, the U.S. Air Force fought hard for the program. The 1990–91 Desert Shield and Desert Storm demonstrated the criticality of air transport during the largest airlift operation in military history. Had the C-17 been available during the Gulf War, the U.S. Air Force could have met its airlift commitment 20 to 35 percent faster. The service estimated that if it had had 80 C-17s instead of 117 C-141s it would have put 50 percent more forces on the ground in Saudi Arabia during the first twelve days of the operation. Also, during Operation Restore Hope in Somalia, the restrictive conditions on the ground at the Mogadishu airport limited the throughput of cargo by the extant airlifters. The C-17, with its superior ground maneuvering capabilities, could have doubled that throughput.

DESIGN REQUIREMENTS

The C-17 is a tremendous advance in airlift aircraft technology. The modern C-17 "glass" cockpit and automatic system monitoring are intended to ease aircrew workload. One important goal was that a crew of just three persons, two pilots and a loadmaster, operated the aircraft. The C-17 is the first large military transport to fly with such a small crew. This requires a high degree of system automation and reliability to eliminate the navigator, flight engineer, or additional loadmasters, and provides a significant reduction in life-cycle cost. The jet is also designed to be operated by small female and very large male crewmembers — a particularly challenging requirement. Despite the use of such advanced technology, the aircraft and its systems were still required to be robust and survivable in the combat environment.

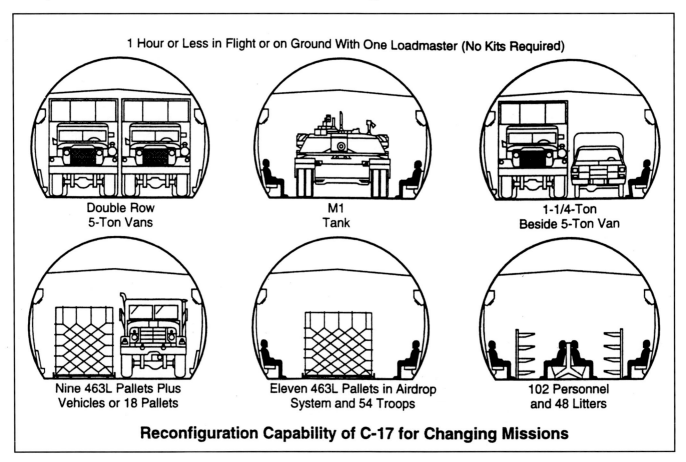

1 Hour or Less in Flight or on Ground With One Loadmaster (No Kits Required)

Double Row
5-Ton Vans

M1
Tank

1-1/4-Ton
Beside 5-Ton Van

Nine 463L Pallets Plus
Vehicles or 18 Pallets

Eleven 463L Pallets in Airdrop
System and 54 Troops

102 Personnel
and 48 Litters

Reconfiguration Capability of C-17 for Changing Missions

The internal volume of the C-17 provides tremendous loading flexibility. These examples show how military vehicles, including an M1 tank, palletized cargo intended for airdrop, personnel, and litter-borne casualties can be carried with ease. (Boeing)

In celebration of the first year of C-17 flight testing, the team put up all four C-17s then in existence. Photographed over Edwards AFB on 14 September 1992, T-1 leads the formation. Note the darker shade of gray paint on P-2, the first aircraft aft of T-1. (McDonnell Douglas)

or astronautics in America, with respect to improving the performance, efficiency, and safety of air or space vehicles, the value of which has been thoroughly demonstrated by actual use during the preceding year." The trophy has been called the greatest and most prized of all aeronautical honors in America. That year, 1994, was the peak year of the C-17 flight test program and the year of initial operational missions. The award citation read: "The Collier Trophy is awarded to the U.S.

Air Force, the McDonnell Douglas Corporation, the U.S. Army, and the C-17 industrial team of subcontractors and suppliers for designing, developing, testing, producing and placing into service the C-17 Globemaster III, whose performance and efficiency make it the most versatile airlift aircraft in aviation history."

The C-17 flight testing at Edwards AFB continues to collect engineering data on enhancements to the aircraft while operational evaluation of sys-

tem changes are conducted at Charleston AFB, South Carolina. Aircraft P-1 through P-5 were updated to production configuration and delivered to AMC. Its nose boom "proboscis" removed, T-1 remained at Edwards and production aircraft are borrowed as required. T-1 will remain a flight test asset until no longer required and then retired. The CTF eventually became part of the 418th Flight Test Squadron.

Three of the Globemaster IIIs participating in the RM&AE operational testing are shown at Barstow-Dagget Airport, California, not far from Edwards AFB. Visible are P-16 (foreground), P-13, and P-5 (background). All wear what became standard Air Mobility Command markings. (Boeing)

This angle on P-52 shows the two white SATCOM antennae on the centerline and the open AR receptacle. The dark outlines are where personnel are permitted to walk atop the jet without risk of damaging the structure or systems. (AFFTC)

This fascinating line-up of Globemaster IIIs at Charleston AFB, South Carolina, was taken at about the time the first squadron of C-17s was declared operational. The 17th Airlift Squadron, 437th Airlift Wing, was the initial U.S. Air Force C-17A unit. Note the outline of the FEDS "hatch" just forward of the wing root atop the forward fuselage. (Boeing)

WARBIRDTECH
SERIES

THE C-17A MISSION

The mission of a modern military airlifter, particularly one with a tactical role, is not as simple or blasé as many might believe. The C-17's advanced systems and capabilities make this even more true for the Globemaster III. From precisely navigating thousands of miles to a specific point on the earth's surface in flights lasting 10 or more hours, carefully coupling with an airborne tanker to take on thousands of pounds of fuel, paratroop or cargo drops exercising complex systems, to short-field landings on dirt surfaces at high gross weights, the C-17 missions are both challenging and exciting. Following such a flight from beginning to end will illustrate this point.

WALKAROUND

The Globemaster III is a giant transport measuring as long as a C-141B and nearly as wide as a C-5. The wing has a pronounced anhedral and carries four big under-slung engines. Strakes installed on the engine nacelles shed a strong vortex at elevated AOA to delay the onset of flow separation on the wing, effectively increasing the maximum lift coefficient.

The aircraft is equipped with four rugged fuselage-mounted main gear assemblies, two mounted in tandem on each side of the fuselage. Each assembly has three wheels (12 total), with two wheels also found on the nose gear. Strakes on either side of the aft fuselage adjacent to the cargo door reduce drag produced by the large up-sweep required to accommodate freight loading.

Cargo is loaded through the rear where a cargo door raises up into the aft fuselage and a ramp is lowered. The door and ramp may be operated independently from various panels and, as a backup manual capability, using a hand pump.

Two paratroop doors are installed forward of the cargo ramp. Personnel typically enter through the crew entry door on the forward fuselage, port side. This is a fold-down hatch with built-in steps and handrail.

Aircraft P-12 is captured on film at Guantanamo Bay, Cuba, on 26 January 1995. Its supplies for U.S. forces on the island and for refugees housed at the American base have already been offloaded. Note the open APU intake door near the top, forward end, of the starboard sponson and how the inboard elevator segments droop without hydraulic pressure. (DoD)

The cargo compartment is so cavernous that it is uncommon for a load to fill the aircraft and also reach the maximum payload weight capacity. At the forward end of the compartment is a loadmaster station with seat (plus a nearby jump seat) and control panels, a lavatory, small galley, stowage space, and access to avionics racks. The flight deck is accessed via a stairway in this area, the top opening covered with a hinged door panel. The flight deck consists of the cockpit and a crew rest area separated by a bulkhead containing electrical and avionics equipment and circuit breaker panels. The crew rest area has two bunks, two aft-facing seats, a small worktable and stowage space. This is separated from the cargo area by a wall that contains a hemispherical window. There are two rail-mounted Additional Crew Member (ACM) seats aft of the pilot and copilot seats for observers or evaluation pilots. The high degree of system automation eliminates the flight engineer and navigator stations seen in earlier generations of airlifters.

FLIGHT PREPARATION

In preparation for flight the last of the outstanding maintenance items are corrected. Almost all of the aircraft systems include built-in test (BIT) features to reduce the requirement for routine aircrew monitoring. The results of many BITs are stored in computer memory for later download by maintenance personnel. This information is retrieved via the laptop computer at the loadmaster's station and greatly facilitates speedy repair. Equipment likely to need attention during sortie generation and quick-turns at en route stops are readily accessible via quick-release panels.

The aircraft is fueled using two delivery hoses attached to the single point refueling (SPR) station in the rear of the starboard gear sponson or pod. A panel at the SPR station allows control of the fuel system pumps and valves for the desired fuel distribution. The Globemaster III has an enormous fuel capacity in excess of 181,000 pounds (over

27,000 gallons). The jet may defuel by pumping the fuel out of the SPR stations to a tanker truck or to another aircraft or vehicle. The fuselage exterior is also provided with an interphone connection for ground personnel to communicate with the aircrew, electrical power, compressed air and avionics cooling air receptacles, and a liquid oxygen filler port.

The C-17's logistics airlift mission includes the transport of palletized cargo, rolling stock, troops, and aeromedical evacuation patients. The C-17's outsized cargo capability is matched with its considerable maximum payload weight. This permits a wide range of equipment to be carried, such as heavy armored vehicles, helicopters, or a double row of cargo pallets. The majority of the vehicles and aircraft may be transported without the partial disassembly required for other airlifters. Among these loading configurations are: four UH-60 helicopters, three AH-64 helicopters, six 5-ton trucks, one main battle tank with two armored personnel carriers,

Aircraft P-39 is seen under tow at Edwards AFB during the annual open house airshow in October 1999. Production aircraft are occasionally borrowed to conduct flight tests at the U.S. Air Force Flight Test Center, supplementing aircraft T-1. Note the open pilot side window. (Bill Norton)

WARBIRD**TECH**
S E R I E S

five-wheeled Light Armored Vehicles, and much more.

The 170,900-pound cargo capacity is less than the C-5's 261,000 pounds, but much greater than the C-141's 90,000-pound payload and the C-130's 50,000 pounds. The Globemaster III can carry 40,000 pounds of freight on its cargo ramp alone. The C-17 difference is that it can carry this wider variety and greater weight payload over intercontinental distances directly to forward-deployed troops while also operating as a tactical airlifter, airdropping its cargo in adverse weather or landing at short, semiprepared airstrips.

Opening the cargo door and ramp creates an enormous opening capable of accommodating the entry of large loads. The ramp is typically lowered level for rapid loading from a standard K-loader, forklift, or truck bed. It is lowered to the ground for roll-on of wheeled vehicles. Four large loading ramp toes provide a smooth transition over the edge of the ramp or between the ramp and loaders. The toes are typically stowed at the edge of the cargo door. With the ramp and door closed the toes are balanced to allow them to be rolled into position by a single loadmaster and attached to the ramp where they may then be hydraulically lowered and raised. Twin struts forward of the ramp may be hydraulically extended to the ground to add additional aircraft stability during loading of heavy vehicles, particularly on uneven surfaces. The struts may also be used to partially raise the aircraft for main gear tire changes without the use of jacks. Numerous lights illuminate the loading area around the ramp at night.

The cargo floor and ramp include eight rows of roller trays. These are stored in the floor with the rollers either positioned up or turned over to provide a smooth floor for wheeled cargo. The rollers allow easy movement of cargo by a loading crew pushing the pallets into place from the loader or by the loadmaster reconfiguring the cargo in-flight. Cargo may also be pulled aboard using a hydraulic winch installed under the floor at the forward end of the compartment. The winch is operated from a panel or remotely using a cordless hand controller and has a nominal 7,500-pound capacity. The capacity is augmented by the use of pulleys for up to 9,000 pounds. Cargo is restrained using chains and straps attaching to any of the 295, 25,000-pound capacity, cargo tie-down rings that fold down flush with the floor.

Rails in the cargo floor, raised using a simple hand tool, serve as guides and restraints for cargo pallets. Two rows of these rails 108 inches (9 feet) apart, straddling the floor and ramp centerline, are known as Aerial Delivery System (ADS) rails. These position a single row of Type V ADS platforms that are 9 feet wide and from 8 to 64 feet long (variable in 2-foot increments). Up to 11 of the smallest ADS platforms may be carried, with two on the ramp. Two sets of rails 88 inches apart are called logistic rails. These position two rows of up to 18 (with 4 on the ramp) 88 x 108-inch 463L pallets. The rails incorporate locks to hold the pallets or platforms in place and to release them for airdrop. Up to two rows of 20 4 x 4-foot Container Delivery System (CDS) pallets may also be carried. These utilize the ADS rails and the inboard pair of logistic rails. Since these simple plywood pallets cannot use the rail locks, a buffer stop barrier of tubing

McChord's P-51 is sprayed with de-icing fluid during a cold stopover at Elmendorf AFB, Alaska. On the centerline may be seen the two SATCOM antennae with a GPS antenna in between. (DoD)

and chains is erected at the forward end of the cargo compartment to provide emergency restraint. Cargo mixing and airdrop of just a portion of the cargo is possible.

The compartment sidewalls include permanent fold-down seats for 54 passengers/paratroopers. A back-to-back row of 48 additional seats (8 six-seat units) may be installed on the centerline of the aircraft, bringing the maximum passenger capacity to 102. Litter stanchions for up to 36 patients, including oxygen and electrical connections, may be installed in the center of the cargo compartment. Only three of these three-litter stan-

chions are routinely carried, these being stowed in the compartment sidewalls. The seats and litter units are designed to be installed in just minutes, the former otherwise stored in cargo door cavities. Up to 100 airline-type seats may be installed by loading 10 ten-seat pallets.

Using the weight and center of gravity data provided for each pallet and vehicle loaded, the loadmaster uses a laptop computer at the forward loadmaster's station for cargo weight computations and CG determination. The pilots also use computers for their mission planning. Voluminous data regarding waypoints, nav-

igation aids, instrument approaches, and so forth — essentially every navaid in the world and about 500 major navigation fixes — are preprogrammed. En route fixes, tanker rendezvous tracks, airdrop points, and other information may be added at the squadron and then carried to the aircraft on a floppy disk. Likewise, communication data may be preprogrammed so that frequencies and appropriate radio selections are readily available in flight.

The preprogrammed mission planning data is loaded quickly through a laptop computer plugged directly into the aircraft's three redundant

Cargo pallets are prepared by an aerial port squadron and then placed on a K-loader. Once positioned behind the ramp of the aircraft, the loader allows rapid transfer of the pallets onto the C-17. Note the gaps above and below the rudder segments to allow for free movement during structural flexing. (DoD)

WARBIRD**TECH**
S E R I E S

A C-17 loadmaster directs a driver of a fuel truck rolling off a C-17 at Skopje Airport, Macedonia, on 17 June 1999 during NATO's Operation Joint Guardian. Airlifting such heavy equipment into small airfields like Skopje is the hallmark of the Globemaster III. (DoD)

mission computers (MCs). Complete missions, including horizontal, vertical, and time profiles, may be preprogrammed for entirely automated in-flight execution by the computers including timed arrival at waypoints, airdrop, landing approaches, and so forth. Automated weight and balance data plus takeoff and landing data computation is available. The MCs are accessed through two keyboards in the cockpit center pedestal with the status or information provided on four cathode ray tube (CRT) displays. The preprogramming feature reduces the time required to ready the aircraft for flight.

DEPARTURE

Pilot training for upgrades to aircraft commander and for various specialty skills is accomplished with weeks of desktop computer-based instruction and simulator time but with only two or three flights at the end of a course. This is possible because of the highly automated nature of the C-17 and its easy handling thanks to the EFCS. The greatest challenge is knowing the complex aircraft systems plus remembering how to access and use all the available flight information. The crews also spend many hours a month in

the simulator at their home station practicing emergency procedures and other drills. This greatly saves on training costs.

When the crew climbs aboard the aircraft the loadmaster sees to seating the passengers. The pilots upload the mission data and take their seats. Typical of transport aircraft, a full width instrument panel is augmented by the center pedestal console and overhead panel between the pilots. Four large full-color multi-function displays (MFDs) with surrounding keys dominate the instrument panel and display the principal flight and engine information. The MFDs may be set up in five basic display formats depending on the phase of flight as well as numerous other "pages" of information. Also serving as a primary flight instrument is a stowable HUD in front of each pilot. Among other instrument panel primary items is the landing gear handle. Above is a shallow glareshield panel. The center pedestal contains the throttle quadrant, plus flap and parking brake handles. Both pilots and each ACM station have adjoining side consoles with accessories and circuit breakers.

The crew starts the four turbofan engines with the auxiliary power unit (APU). The APU is housed in the forward portion of the starboard landing gear sponson, the pod consequently extending forward farther than the port sponson. The unit provides electrical power and compresses air in addition to facilitating engine starting without the use of ground carts. The APU air intake door is mounted above the large circular exhaust, both forward of the APU. An alert-starting feature allows the APU to come online automatically in about a minute to prepare for immediate engine start after the pilots take their seats. The automatic cycle is initiated by pressing a button just inside the crew entrance door. APU starting power is from a ground cart or aircraft batteries.

One of the most demanding requirements for the C-17 was delivery of an M1 Abrams main battle tank into a semiprepared field near the combat zone. Here a U.S. Marine Corps M1A1 is backed into P-1 during loading trials at Edwards AFB in early February 1993. Note the lowered stabilization strut on the left, just forward of the ramp, and the ramp toes providing a smooth transition to the ground. (AFFTC)

For a night mission the pilots use the landing lights that fold down out of the wing tips and landing/taxi lights in the forward fuselage to light the taxiway ahead and employ lights in the front of the gear sponsons to illuminate turns to the side. Moderate nose gear steering angles are commanded via the rudder. For low-speed taxi operations, aircraft hand steering tillers, one within reach of each pilot on the cockpit sidewalls, permit greater steering angles. Take-off distance at maximum gross weight is 7,740 feet with the normal engine-out safety margins, but a takeoff run of just 3,000 feet is possible with 160,000 pounds of cargo.

En Route

If the C-17 is carrying a heavy combat load to a distant combat zone it will normally require air-to-air refueling en route. The KC-135 or KC-10 tanker will likely be holding in an orbit in a prebriefed track. As the C-17 closes on the location the pilots initially look for the tanker using the I-band radar beacon transponder navigation radio, with relative position of the tanker shown as a symbol on one of the MFDs. The tactical air navigation (TACAN) radio can also give them the air-to-air range and bearing. As they close on the tanker the Globemaster III pilots establish voice radio contact and complete the rendezvous visually. They prepare

Here three M2 Bradley Fighting Vehicles have been loaded and chained down inside a C-17. The plywood shoring is to prevent the tracks of the armored vehicles from tearing up the no-skid material on the floor. Such vehicles are backed into the jet so they may quickly roll out at the destination where the C-17 may be required to depart rapidly to make room for the next transport or to avoid enemy fire. (AFFTC)

for refueling by opening the Universal Aerial Refueling Receptacle Slipway Installation (UARRSI) located on the upper fuselage centerline, just aft of the flight deck, using a handle on the aft face of the center pedestal. They also reach overhead to configure the fuel system valves and pumps to send the fuel they will take on to the desired tanks. Aerial refueling system status is displayed via lights on the windscreen center post.

The pilot carefully maneuvers into contact position behind the tanker and the boom operator guides the flying boom into the UARRSI. The pilots hear a "clunk" overhead as the boom nozzle seats into the coupler, as verified by the status lights. The fuel panel shows that fuel is being pumped onboard. The pilot is making small stick and throttle changes to hold the aircraft in position. As the desired amount of fuel is taken on the pilots call for a disconnect and the boomer obliges. Alternately, the pilots may initiate disconnect with switches on the sticks or on the throttles.

As the crew approaches the general area of engagement they rendezvous with several other C-17s. These form up as pairs in an approximate line-astern formation with a 2,000-foot separation. This is just one element, or "slice," of several formations that will make airdrops or assault landings at the drop zone (DZ). For night operations the pilots also use formation lighting to help hold position.

Another impressive load-out of the C-17 shows M977 and M998 "Humvee" vehicles in the foreground, an M110A2 203mm self-propelled howitzer in the middle, and an M35A2 truck and M105A2 ammunition transporter at the front of the cargo compartment. Note that the Humvees were actually secured to the ramp before it was raised and locked into position. (AFFTC)

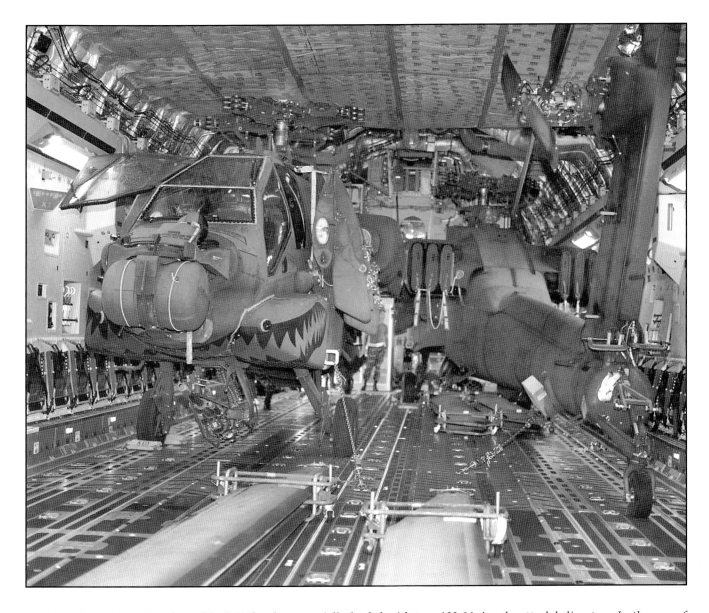

An aircraft carrying aircraft — this C-17 has been partially loaded with two AH-64 Apache attack helicopters. In the case of these helicopters, the rotor blades have been removed for transport aboard the Globemaster III. (Boeing)

Formation lighting consists of 10 white lights at various locations on the aircraft and eight large green electroluminescent panels on the fuselage sides, winglets, and vertical stabilizer.

If the formation flight is made more challenging by light rain showers and patchy clouds at the flight altitude, the crews have an electronic aid called the Station Keeping Equipment (SKE). The lead aircraft's SKE system transmits signals to the other aircraft in the formation, driving the guidance cues on the flight director page of the MFD for each pilot to maintain position, initiate turns, execute airdrops, and so forth. The participating aircraft also transmit signals to show their relative position in the formation as displayed on the SKE plan position indicator MFD page. The autopilot (AP) and autothrottle system of a follower aircraft may also be slaved to the SKE guidance signal through the mission computer to automatically keep position. Thus, with the lead aircraft flying on AP and following a course preprogrammed into the mission computers, and the rest of the formation on AP, a formation of up to 18 C-17s can fly in formation in weather and execute an airdrop "hands-off."

The leader's weather radar is used by the crew to find their way around the worst of the weather cells. When switched to ground mapping mode the radar also helps to verify position. The jet's comprehensive navigation system, with four redundant inertial reference units and two

Opening (HALO) paratroop jumps above 12,500 feet. Personnel airdrops are normally performed at 500 to 800 feet and at 130 to 145 knots.

The Globemaster III's aerial delivery systems provide the capability for airdrop of 102 paratroops or, employing recovery parachutes, palletized loads may be airdropped. The C-17 is the only aircraft (save for a rarely exercised C-5 capability) that can airdrop outsized cargo. The C-17's ability to airlift heavy armored vehicles such as a main battle tank, a self-propelled heavy gun, or a multiple rocket launcher directly to forward battle areas represents a tremendous combat enhancement for the U.S. armed forces. Aircrews have referred to the C-17 as the "four-wheel drive airlifter" because of its tremendous tactical utility.

The "bread-n-butter" of U.S. Air Force airlift is the high-volume transport of logistics supplies. This image shows a partial load of palletized cargo. Plastic wrap and cargo netting keep the articles together during transit to and from the aircraft and during flight. (Boeing)

Global Positioning System (GPS) receivers, can suitably guide the crew to the DZ. For airdrops in weather, a drop zone point of impact is usually identified using the SKE to display the location of an AN/TPN-27 zone marker previously positioned on the ground by a special forces team.

As the C-17 approaches the area of enemy activity, the pilots descend to as low as 300 feet so that they will be difficult to observe on enemy radar. They also shut off external lights and configure the cockpit for Night Vision Goggle (NVG) use. Only the pilot flying dons the NVGs that allow him to see the world outside in shades of green. The landing/taxi lights may be fitted with infrared (IR) filters for use with NVGs to

reduce the likelihood of the aircraft being visually tracked at night. When deployed, these lights may be positioned as required to illuminate an LZ. The crew also activates the defensive countermeasures system that will automatically eject flares should the equipment detect a missile launch in the area of the aircraft.

AIRDROP

The MC can automatically determine the Computed Air Release Point (CARP) and indicate the optimum instant for the drop sequence to commence. With the cargo door and ramp open the aircraft can be operated at up to 25,000 feet altitude and 250 knots (0.6 Mach). This capability would normally only be used for depressurized High Altitude - Low

The U.S. Army's strategic brigade airdrop, vital for meeting the two nearly simultaneous major regional conflicts, is a mission virtually made for the C-17. The brigade is to be delivered anywhere in the world with 48-hour notice. It requires airlift from CONUS directly to the combat zone for both airdrop and airland (troop insertion via landed aircraft) at an austere airfield. The mission, at the time the C-17 demonstrated its abilities, called for a 30-minute airdrop of 2,552 paratroopers and 1,350 tons of equipment, including 116 wheeled vehicles, 10 20-ton Sheridan light tanks (since deleted with retirement of this vehicle), 18 105mm howitzers, and 54 equipment bundles. Within 24 hours additional materiel would be delivered via airland at a captured austere airfield, consisting of 690 more troops, 224 wheeled vehicles, 28 helicopters, and 33 equipment and supply bundles. The second phase alone would require 43 C-17s.

In preparation for the troop drop the loadmaster opens the paratroop doors. Since the lower half of the troop doors falls within the aft portion of the main gear sponsons, each door has an attached fairing that retracts as the door is pulled in and slid up. With the paratroop door open, an illuminated jump platform or "porch" is exposed within the gear pod just outside the door. Ahead of each door is a perforated air deflector that is deployed to provide optimum paratroop jump conditions, allowing the paratroopers to exit the aircraft with less difficulty. The troop doors are larger than those on other airlifters, reducing the chance of injury on paratroop exit.

Red, amber (15 second warning), and green jump lights adjacent to the door are similar to those used since World War II. Jumpers attach their static lines to overhead anchor cables strung along both sides of the cargo compartment. A jumper whose static line fails to release would trail behind the aircraft. The trooper may then be pulled aboard using a static line retriever winch. The winch is normally used to pull in the static lines and parachute deployment bags ("D" bags used to hold the canopy and risers) that collect outside the aircraft. Small parachute-recovery bundles weighing up to 500 pounds each may also be pushed out the troop doors in the same manner as paratroopers.

Cargo drops via the ramp are normally made from a 650 to 1,150-foot altitude. Drogue parachutes, for deployment of the main cargo extraction chutes, are stored in recessed housings on either side of the cargo door opening for redundancy or multiple pass airdrops. They are deployed hydraulically by an electrical signal from the loadmaster or the pilot, although a manual backup is provided. Once deployed, the drogue inflates in the slipstream but remains attached to the aircraft via a tow release mechanism at the end of the ramp. Should the decision be made to abort the drop, the tow release allows the drogue to be jettisoned. A backup

Other common "cargo" are troops being deployed on exercises, in reaction to international emergencies, and to relieve those already in place. These paratroopers load-up for a drop during the RM&AE testing. Visible are P-11, P-5 with "TEAM AFOTEC" tail markings, and Edwards AFB's P-1 with special grid pattern painted on the aft fuselage to facilitate paratroop tests. (DoD)

This loadmaster pulls the pins on the nose gear preparatory to departure. His headset is connected via a long cord to the interphone connection inside the crew door on the port side of the aircraft. The access panel on the nose is for servicing of the passenger oxygen supply. (DoD)

guillotine provides an emergency drogue line cut option. A small video camera is installed within the end of the ramp to allow the load-master to verify the condition of the drogue and extraction chutes on a small monitor at the forward load-master station. An IR lamp is also installed in the ramp to illuminate the chute for viewing in darkness.

When the pilot or loadmaster activates the release the drogue is deployed and pulls out the main extraction parachutes. When the floor cargo rail locks are loaded to a specified force, upon extraction parachute inflation, the locks release automatically. Extraction chutes of 15- to 35-foot diameter may be used singly or in clusters. These pull the load from the aircraft and also deploy the main recovery parachutes to permit a safe descent to earth. Alternatively, gravity drop, employing up to 2,350-pound CDS pallets, may be employed by releasing the cargo restraints with the aircraft flying at a high deck angle. As the pallets slide off the ramp, static lines pull a pilot chute out and, in turn, the recovery parachutes.

Low Velocity Air Delivery (LVAD) of ADS pallets are employed for large and heavy items and decelerated with as many as 10 parachutes. Up to 110,000 pounds may be dropped via LVAD with a single Type V platform weighing as much as 60,000 pounds. However, the U.S. Army lacks anything approaching this weight, the largest items being bulldozers and scrappers used to prepare airfields. Only 8 of the 11 smallest ADS pallets can be dropped (excluding those on the ramp and the last one forward) using the parachute extraction method. As many as 40 CDS pallets may be dropped via the gravity method in one or two

rows ("sticks") of 20. Each may weigh as much as 2,350 pounds and descends under a single parachute. For a CDS drop, restraining straps are released electrically.

The C-17 was also designed to deliver up to 60,000 pounds via low altitude parachute extraction. This involves flying 12 feet or less off the DZ while loads are extracted via parachute to decelerate while sliding along the ground. The LAPES capability was demonstrated up to 42,000 pounds but further testing was terminated when the U.S. Army decided in 1994 to discontinue the mission. The LAPES training proved hazardous and costly over the years and the capability had not been used in combat since the Vietnam War.

STOL OPERATIONS

Many aspects of the jet make STOL operations possible. Direct lift control facilitates the steep approach and precise flight path control by use of a thumb switch on the outboard throttle levers to command spoiler movement. Regulating the spoiler deflections via the thumb switch allows more rapid variations in flight path than the large, slow-responding turbofan engines can provide. A throttle-spoiler interconnect bias (TSIB) also allows the spoilers to augment the response to throttle movement for rapid airspeed and flight path control. The spoilers are biased up 6 degrees at 3/4 flaps and 9 degrees at full flaps. They retract to produce a rapid decrease in flight path angle or extend up to 25 degrees to produce a rapid increase in flight path angle. The spoiler response is automatically adjusted for optimum effect at various flight conditions. The spoilers also augment aileron roll authority at low airspeeds.

The five-degree steep approach is instrumental in making short-field assault landings. Instead of the usual 750-foot long touchdown zone, the steep approach allows the C-17 to routinely land within 200 feet of the aim point. The precise flight path guidance for STOL approaches using the backside technique requires augmentation of the basic aircraft stability and handling qualities via the EFCS and advanced cockpit display aids. The approach flight path vector is displayed on the HUD, allowing the pilots to keep the flight path vector superimposed on the intended aimpoint for precise touchdown control. The large lower "downview" windows at the pilots' feet are critical for visual identification of ground references and the LZ during approach. The aircraft can also change airspeed quickly by using speed brakes and thrust reversers when tactical situations demand. These features permit the aircraft to spiral down directly over the LZ with a tremendous descent rate before setting up for landing, reducing exposure to ground fire.

Ground effect (additional lift produced in proximity to the ground) is augmented by the blown flaps with application of a throttle-push at the appropriate radar altitude (using the two Combined Altitude Radar Altimeters). This adds to the flap jet and, because of TSIB, causes the spoilers to retract and provide excess lift. This technique cushions the landing and allows the aircraft to set down without the need to flare, further reducing touchdown dispersion and eliminating the likelihood of ballooning or a hard impact. These features allow the C-17 to land at about 110 to 120 knots, approximately the same as a C-130, and to reduce the landing rollout distance. Once firmly on the ground the Automatic Ground Spoilers command all spoilers to move symmetrically to 60 degrees deflection to increase drag, reduce lift, and improve wheel brake efficiency to

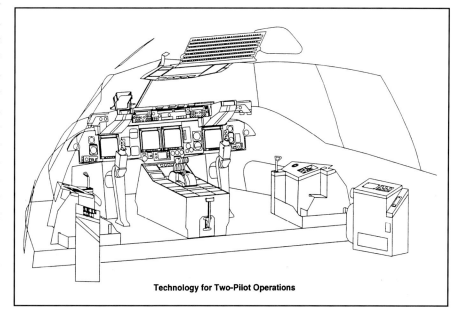

Technology for Two-Pilot Operations

The two-person cockpit is highly automated. The arrangement is shown here with the two pilot seats and two Additional Crew Member (ACM) seats removed. Consoles beside the ACM seats contain little more than intercom panels and oxygen regulators. (Boeing)

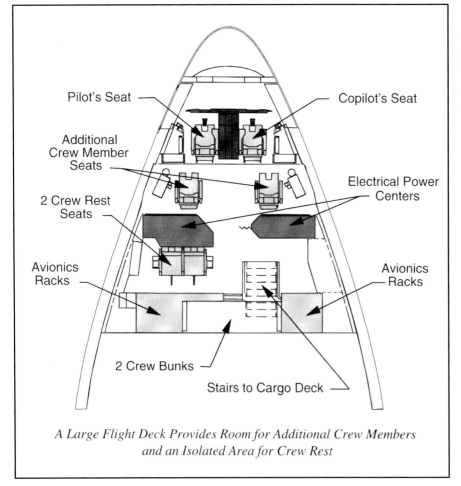

Pilot's Seat

Copilot's Seat

Additional Crew Member Seats

2 Crew Rest Seats

Electrical Power Centers

Avionics Racks

Avionics Racks

2 Crew Bunks

Stairs to Cargo Deck

A Large Flight Deck Provides Room for Additional Crew Members and an Isolated Area for Crew Rest

The flight deck provides maintenance access to electronic buses and power equipment in the Electrical Power Centers that separate the cockpit from the crew rest area. Apart from providing the stairs down to the cargo compartment, the crew rest area also contains two aft-facing seats, two bunks, and maintenance access to avionics racks. Crew flight gear is commonly stored to the right of the stairs. (Boeing)

assist in deceleration. Engine thrust reversers and high-energy wheel brakes allow the aircraft to stop in a short distance.

High sink rate landings, pivot turns at high gross weights, operations on semiprepared fields, and heavy braking for STOL landings all create a demanding landing gear load spectrum. The gear is designed to withstand a 12.5 feet per second landing impact, nearly comparable to the sink rate for normal aircraft carrier landings. The large multiple tires provide the aircraft with a

remarkably light footprint to allow soft field operations. The wheels are each equipped with a stack of six carbon brake disks. An electronic anti-skid system automatically prevents locking of the brakes while permitting maximum effort braking for rapid deceleration.

Retarding the throttles to idle, lifting them up, and pulling them aft deploys engine thrust reversers. Moving the throttle farther aft will command greater reverse thrust, although only idle reverse thrust is selectable in flight. Aft portions of

the fan and core cowling translate aft to expose cascade grills. Simultaneously, blocker doors in the fan and core sections block and redirect the annular airflow through the cascades.

GROUND OPERATIONS

A small austere airfield, captured or created in an air assault, may have very little room for the aircraft to move about. With other aircraft waiting to land the C-17 must be unloaded quickly and then depart. The crew will keep the engines running but with the thrust reversers deployed. The thrust reversers direct exhaust upward and forward. This reduces the likelihood of ground debris and dust being kicked up, obstructing vision for ground operations and possibly creating a Foreign Object Damage (FOD) hazard to the engines. Offloading and loading with engines running can be rapidly accomplished without the fear of jet blast and without the time required to restart the engines, permitting a quick turn-around. The compact aircraft dimensions, tight turning radius, and ability to back up using reverse thrust allows ground operations at fields with minimal ramp space under congested conditions. Up to six C-17s can maneuver and park on a 300,000 square foot ramp as compared with two C-141s or a single C-5.

If a K-loader or similar ground vehicle is present, the jet may be offloaded in just minutes even with a full load of 18 large 10,000-pound 463L pallets. To help facilitate a rapid unloading and departure under combat conditions, the C-17 may deposit its cargo using the combat offload technique. This involves releasing the load restraints as the aircraft rolls sharply forward, caus-

COMBINING AIRLIFT ROLES WITH STOL

ing the load to roll out the back. The short drop off the ramp to the dirt does not harm the cargo and the C-17 taxis away from the line of pallets. The C-17 can deliver a double row of 18 pallets, up to 9,400 pounds each, or 11 10,350-pound pallets in a single row, through this means in just seconds. With the cargo compartment now free, the loadmaster may quickly erect aeromedical stanchions as wounded are brought in on litters and ambulatory patients are guided to seats.

With the aircraft now lightened, the STOL performance of the C-17 is important in getting out of the short field and quickly climbing away from possible ground fire. The pilot sets partial flaps, runs the engines up while holding the brakes and then, with a jolt, releases the brakes and begins a rapidly accelerating ground roll. In less than 2,000 feet the pilot smartly rotates the aircraft up and the big transport leaps off the ground in a steep climb.

The C-17 leaves the area the way it came, at low level, until it is safe to climb to a more normal operating altitude. There it may again take on fuel from a tanker before flying to a friendly base. After landing the aircraft might be met by ambulances to take off the wounded soldiers. The cargo ramp includes an emergency blow-down feature, lowering it in 5 seconds versus the usual 20 so that critical patients may be evacuated quickly. Maintainers, ground support equipment, and war reserve maintenance kits (containing tools and commonly used parts) have already been prepositioned at the base in anticipation of the Globemaster III's arrival. The next day the C-17 will likely be flying in additional rations, ammunition, and other gear for the troops.

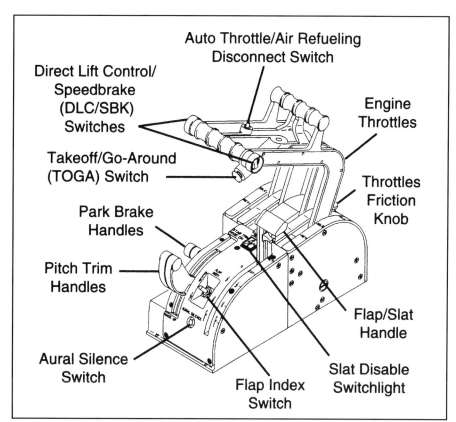

The multiple controls on the throttle quadrant are detailed here. With the throttles aft at idle, they may be lifted up via the top grips and pulled aft to deploy the engine thrust reversers. Moving the throttles farther aft will command greater reverse thrust, although only idle power in reverse is permitted while in flight. (Boeing)

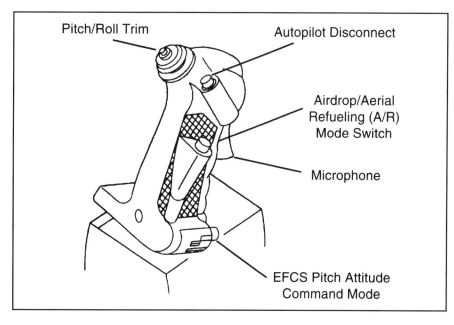

The C-17 is unusual in that it possesses a control stick versus a yoke. While the entire control column moves fore and aft, only the upper portion with the grip moves laterally. The left seat is provided with a left-handed grip and the right seat with a right-handed grip. (Boeing)

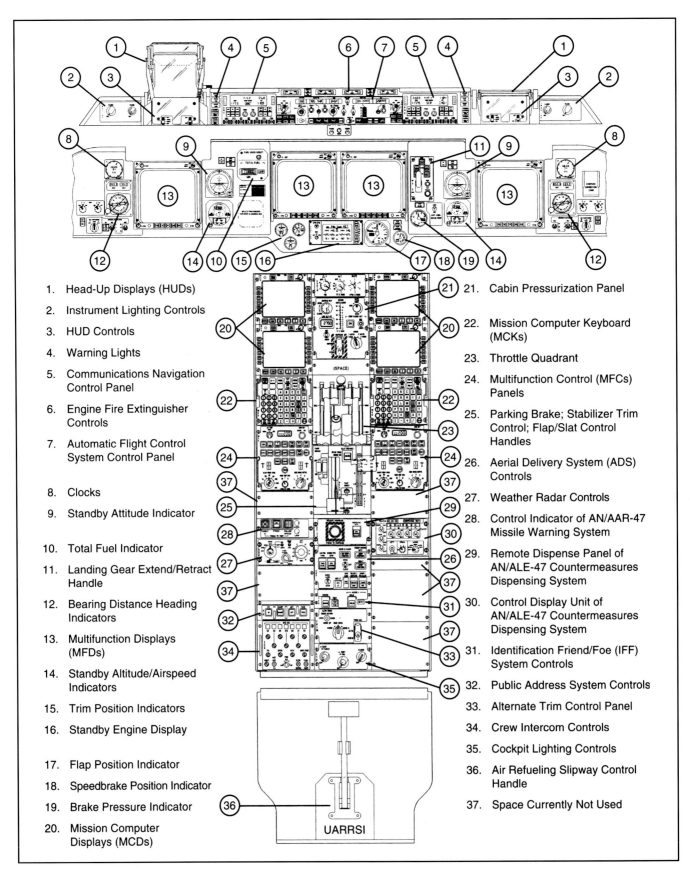

1. Head-Up Displays (HUDs)
2. Instrument Lighting Controls
3. HUD Controls
4. Warning Lights
5. Communications Navigation Control Panel
6. Engine Fire Extinguisher Controls
7. Automatic Flight Control System Control Panel
8. Clocks
9. Standby Attitude Indicator
10. Total Fuel Indicator
11. Landing Gear Extend/Retract Handle
12. Bearing Distance Heading Indicators
13. Multifunction Displays (MFDs)
14. Standby Altitude/Airspeed Indicators
15. Trim Position Indicators
16. Standby Engine Display
17. Flap Position Indicator
18. Speedbrake Position Indicator
19. Brake Pressure Indicator
20. Mission Computer Displays (MCDs)

21. Cabin Pressurization Panel
22. Mission Computer Keyboard (MCKs)
23. Throttle Quadrant
24. Multifunction Control (MFCs) Panels
25. Parking Brake; Stabilizer Trim Control; Flap/Slat Control Handles
26. Aerial Delivery System (ADS) Controls
27. Weather Radar Controls
28. Control Indicator of AN/AAR-47 Missile Warning System
29. Remote Dispense Panel of AN/ALE-47 Countermeasures Dispensing System
30. Control Display Unit of AN/ALE-47 Countermeasures Dispensing System
31. Identification Friend/Foe (IFF) System Controls
32. Public Address System Controls
33. Alternate Trim Control Panel
34. Crew Intercom Controls
35. Cockpit Lighting Controls
36. Air Refueling Slipway Control Handle
37. Space Currently Not Used

The arrangement of instrument panel, glareshield panel, and center pedestal are shown in this illustration of the present configuration. Some items will likely be moved and additional equipment installed over the years of operation. (Boeing)

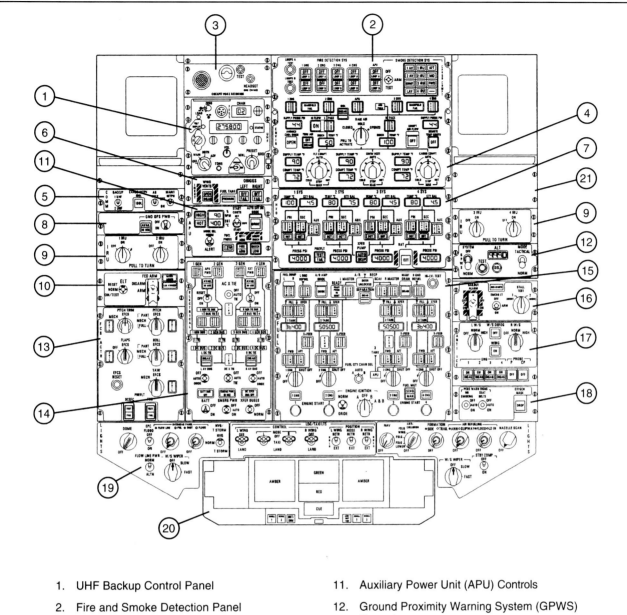

1. UHF Backup Control Panel

2. Fire and Smoke Detection Panel

3. Crew Voice Recorder Controls

4. Environmental Control Panel

5. VHF Backup Control Panel

6. Onboard Inert Gas Generating System Controls

7. Hydraulic Systems Panel

8. Ground Power Control Panel

9. Inertial Reference Unit (IRU) Controls

10. Flotation Equipment Deployment System (FEDS) Controls

11. Auxiliary Power Unit (APU) Controls

12. Ground Proximity Warning System (GPWS) Controls

13. Flight Control System Actuator Panel

14. Electrical System Control Panel

15. Fuel System Panel

16. Bailout Alarm Control Panel

17. Anti-Icing System Controls

18. Passenger Warning System Controls

19. Interior and Exterior Lighting Controls

20. Warning Annunciator Panel

21. Space Currently Not Used

The overhead panel is detailed here. At the lower end is the warning and caution annunciation panel that provides a scrollable display of color-coded messages on three LED screens. (Boeing)

Seen from the boom operator's position of a KC-135, a C-17 closes to the contact position to take on fuel far above the Atlantic. The lines used by the operator to guide the boom to the aerial refueling receptacle are visible atop the nose of the C-17. (DoD)

A rare line-up of four C-17 test aircraft at Edwards AFB is photographed as the jets wait to take the huge runway at the desert base. (McDonnell Douglas)

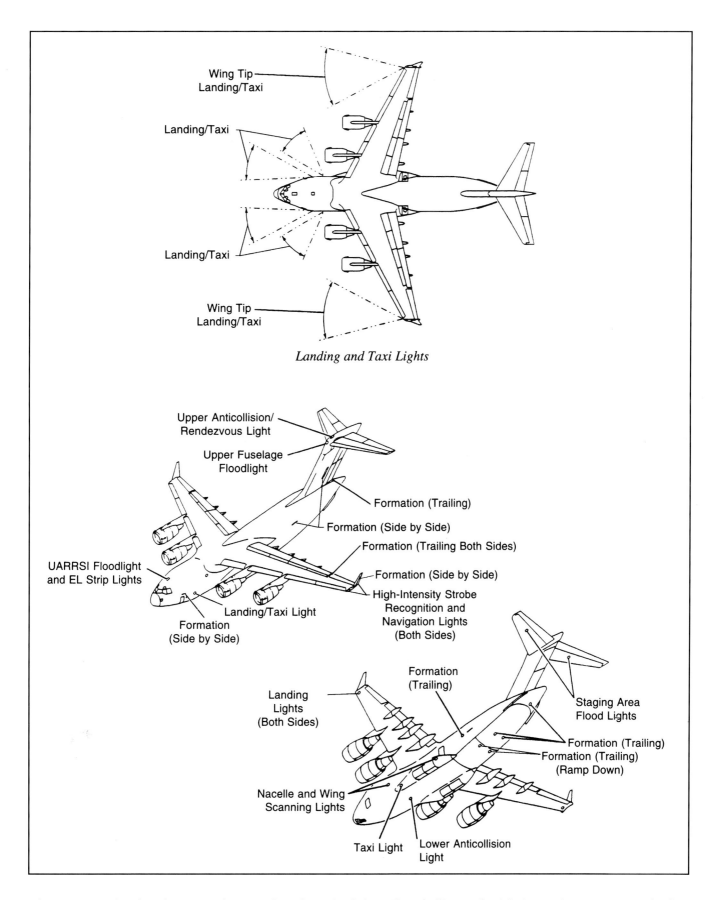

Wing Tip
Landing/Taxi

Landing/Taxi

Landing/Taxi

Wing Tip
Landing/Taxi

Landing and Taxi Lights

Upper Anticollision/
Rendezvous Light

Upper Fuselage
Floodlight

Formation (Trailing)

Formation (Side by Side)

Formation (Trailing Both Sides)

Formation (Side by Side)

UARRSI Floodlight
and EL Strip Lights

High-Intensity Strobe
Recognition and
Navigation Lights
(Both Sides)

Landing/Taxi Light

Formation
(Side by Side)

Formation
(Trailing)

Landing
Lights
(Both Sides)

Staging Area
Flood Lights

Formation (Trailing)

Formation (Trailing)
(Ramp Down)

Nacelle and Wing
Scanning Lights

Taxi Light

Lower Anticollision
Light

The C-17 is equipped with a tremendous number of exterior lights. These facilitate safe nighttime maintenance, cargo loading operations, taxi in confined spaces, and formation flight. (Boeing)

A pair of Globemaster IIIs fly a trail formation as they approach North Field drop zone, near Charleston AFB, in preparation for a practice airdrop. Note that the cargo ramps of both aircraft have already been lowered. Such formations are typically flown in a 2,000-foot line-astern trail at 650 to 1,150 feet altitude. (Boeing)

The last of a plane-load of paratroopers depart out the port troop door of this C-17. Note their static lines attached to the overhead anchor cable and the lines collecting at the door from those who have already jumped. (DoD)

A pair of Globemaster IIIs deposit their load of paratroops over a drop zone at Ft. Bragg, North Carolina. The ability of each C-17 to drop 102 paratroopers with great accuracy at night and during inclement weather is a significant military advantage for the United States. (Boeing)

A string of CDS bundles spills out the back of P-1 during airdrop testing. The drogue parachutes can be seen deploying the descent chutes. An astonishing quantity of supplies can be rapidly and accurately delivered via this means. (Boeing)

The U.S. Air Force began testing the Low Altitude Parachute Extraction System (LAPES) capability of the C-17 over Rogers Dry Lakebed at Edwards AFB. The work was terminated when the military discontinued the LAPES requirement. Here a test load is extracted by two parachutes as the jet flies slowly just a few feet off the ground. (Boeing)

High Altitude-Low Opening (HALO) troops prep for a jump from the Globemaster III. They will pre-breath from the aircraft's oxygen via the hoses seen attached to the sidewall connectors. This also allows them to top off the oxygen tanks they carry with them in the jump. (AFFTC)

C-17A aircraft P-10 flies over the beautiful South Carolina countryside near Charleston AFB. The low pressure over the wing has produced condensation in the moisture-laden air. (DoD)

These loadmasters have positioned a bundle of ammunition containers in the port troop door, awaiting the final green light to shove it out. The parachute bag is seen atop the bundle with its static line attached to the sidewall anchor cable. Note the omni-directional rollers in the floor tray that make moving such objects about an easy task. (AFFTC)

A dual row of equipment pallets are positioned on the cargo ramp and floor. The static lines for the drogue chutes can be seen attached to the sidewall cables, ready to deploy the twin recovery parachutes. All that is needed is the proper deck angle and the electrical command to release the restraint locks for the pallets to begin rolling out the back of the aircraft. (USAF)

Aircraft P-1 demonstrates the combat offload capability of the C-17 during a test at the U.S. Air Force Flight Test Center. The restraint locks on the cargo pallets are released and the aircraft abruptly accelerates forward. The cargo then simply slides out of the jet and falls the short distance to the ground, allowing for rapid unloading in conditions where enemy fire prevents normal cargo handling operations. (Boeing)

An extraction parachute pulls a Low Velocity Aerial Delivery (LVAD) load, a 105mm howitzer and ammunition, from a C-17. An instant later the chute will pull the bags off of the recovery parachutes stowed on top of the load. The corrugated cardboard under the gun helps to dissipate the landing impact load as they are crushed. (John Norton)

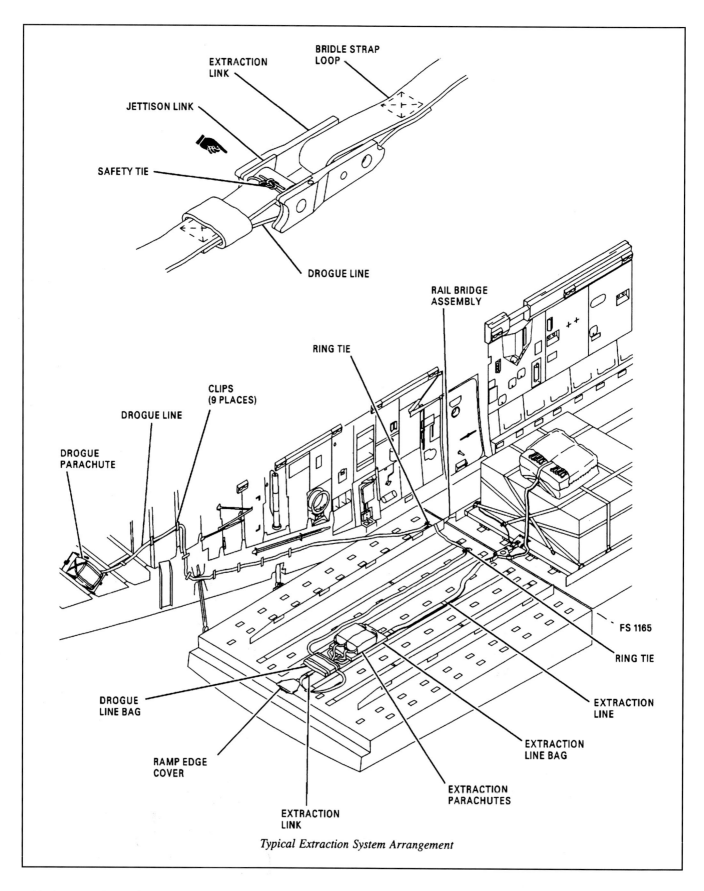

EXTRACTION LINK

BRIDLE STRAP LOOP

JETTISON LINK

SAFETY TIE

DROGUE LINE

RAIL BRIDGE ASSEMBLY

RING TIE

CLIPS (9 PLACES)

DROGUE LINE

DROGUE PARACHUTE

FS 1165

RING TIE

EXTRACTION LINE

DROGUE LINE BAG

EXTRACTION LINE BAG

RAMP EDGE COVER

EXTRACTION PARACHUTES

EXTRACTION LINK

Typical Extraction System Arrangement

This drawing shows the typical LVAD extraction system arrangement. On the right is the load on an ADS pallet positioned in the centerline ADS rails. Atop the load is the recovery parachute. (USAF)

The tires and engines kick up a huge dust cloud behind this C-17 as it begins its takeoff roll from Bicycle Lake Dry Lakebed, near Edwards AFB. The ability of such a large aircraft to operate from semiprepared surfaces in close proximity to a combat zone allows rapid insertion of heavy military equipment and personnel under a broad spectrum of conditions. (DoD)

The C-17 fleet is typically scattered to the four winds, with elements deployed worldwide. This image was captured at Charleston AFB, South Carolina, during Christmas 1999 when the relatively quiet international situation allowed most of the crews and aircraft to remain home. (Bill Norton)

A pair of Globemaster IIIs out of Ramstein AB unload U.S. Army vehicles, part of Task Force Hawk, at Rinas Airport, Tirana, Albania on 18 April 1999. Other C-17s flew into Rinas with humanitarian assistance supplies as part of Operation Sustain Hope, helping to reduce the human suffering created by the refugee exodus from war-torn Kosovo. Note the poor ramp conditions and deep mud that characterized Tirana at the time, indicative of the fields in which the C-17 crews can find themselves. (DoD)

The crew of the first C-17 flight
From left are flight test engineer Henry Van De Graaf; pilot Bill Casey, co-pilot Lt. Col. George London; and loadmaster Ted Venturini

The crew pose in front of T-1 in Long Beach prior to the successful first flight of the C-17. Note the open nacelle doors and the engine covers with the C-17 logo, used only during the test program (Boeing)

C-17 aircrew receive their Globemaster III training at the 97th Air Mobility Wing (AMW), Altus AFB, Oklahoma. One of the unit's eight "flying classrooms" shows off its colorful tail markings. Note the Air Education and Training Command (AETC) label beneath the American flag and a unique moose nose marking. (Boeing)

Air Force Reserve squadrons operate C-17s beside their active duty counterparts in the 437th and 62nd AWs. One Air National Guard (ANG) airlift squadron (AS), the 183rd AS based in Jackson, Mississippi, will receive Globemasters III beginning in 2004. This retouched photo shows the markings the aircraft will bear. (Boeing)

U.S. Marines help get P-4 ready for departure from the now defunct El Toro Marine Corps Air Station, California. Photographed in December 1994 while still a flight test asset, the jet had yet to have its full complement of operational markings applied. Note the open co-pilot side window that also doubles as an emergency exit. (DoD)

To fit inside the C-17, the U.S. Marine Corp's CH-46 Sea Knight helicopter requires some disassembly, as seen here. (Boeing)

GRAY IS BEAUTIFUL

UTILITARIAN IN FINISH AND OPERATION

Aircraft T-1 was delivered with a three-tone camouflage paint similar to a standard European One scheme. This consisted of gray, forest green, and olive green. P-2 was painted in a uniform dark gray. P-1 and subsequent production aircraft were given what is the standard C-17A paint scheme — a uniform shade of medium gray that is applied to all U.S. Air Force airlifters and initially referred to as "Proud MAC Gray." (T-1 was repainted in the standard C-17 fleet colors in the summer of 1999, still bearing the colorful "C-17" logo on the nose and "412TW" under the serial number.)

On P-2 and subsequent aircraft, the engine inlet lips are a bright metal color and the areas around and directly forward of the APU exhaust are unpainted metal panels. With the addition of titanium panels on the leading edges on the slat segments inboard of each engine, with associated paint adhesion issues, these panels were left unpainted. The forward portion of the flaps on the lower surface aft of the engines is a dark metal color. The UARRSI receptacle strike plate is also left natural metal.

Each airlift wing displays a distinctive tail marking between the two black stripes that bisect the tail. Some of the aircraft have been given commemorative "Spirit of ..." names that appear on the forward port side of the fuselage.

C-17A Globemaster III aircraft P-7 shows off the purposeful lines of the most sophisticated and capable airlifter in the world. (DoD)

Aircraft P-55 departs Long Beach in late 1999 for its delivery flight to the 62nd Airlift Wing, McChord AFB, Washington. The 62nd AW was the second operational C-17 wing to get jets after Charleston AFB's 437th AW. The unit's unique tail marking, depicting Mt. Rainier, is a sharp contrast to the uniform medium gray paint that adorns all C-17s. (Boeing)

The sole prototype C-17, designated T-1, dumps fuel over the Mojave Desert near Edwards AFB. The aircraft carried the outdated European One camouflage until 1999. Note also the "C-17" logo on the forward fuselage created with the American flag's "Stars and Stripes." (AFFTC)

One of the flight test birds is seen operating on a dirt field in the southwestern United States. The engine's propensity for generating an inlet vortex during reverse thrust is evident as dirt is drawn into the engine and blown out the reverser cascades. (AFFTC)

The C-17 has already performed its share of emergency relief airlift missions. Here P-18 delivers fresh water and food to St. Thomas, the U.S. Virgin Islands, following Hurricane Marilyn in September 1995. The EFCS yaw damper feature is moving the lower rudder segment during taxi. (DoD)

WARBIRD**TECH**
SERIES

The Air Force never sleeps, as demonstrated by these Charleston AFB C-17s being prepared for another mission during third shift. The nose markings testify to the fact that the 437th AW's reserve counterpart is the 315th Airlift Wing. Note the exposed engine reverser cascades on P-12 in the background. (DoD)

This image illustrates the modernity of the C-17, with its "glass" cockpit, HUDs, and fine accouterments. Hanging on the side window posts are quick-don oxygen masks, at hand in the event of depressurization. The photo was taken during a May 1996 exercise with a flight to Kuwait. (DoD)

Aircraft P-7 was the testbed for the Airlift Defensive System. One test included the ripple firing of its full complement of infrared flares from its four dispensers. The installation is one of many upgrades and changes made to the C-17 during its production. (Boeing)

A Globemaster III is shown on the Long Beach, California, production line after the joining of the wing and fuselage. The engine stub pylons are visible, as is the fitting on the tail used during hoisting via overhead crane. The hoses entering the wing access points deliver fresh air to personnel working within the structure. (McDonnell Douglas)

C-17 AIRFRAME SUPPLIERS

CODE	SUPPLIER
(1)	AEROSTRUCTURES HAMBLE, LTD.
(2)	ALLIED-AIRCRAFT LANDING SYS.
(3)	ALLIED SIGNAL AEROSPACE
(4)	AUTO AIR COMPOSITES
(5)	B. F. GOODRICH/CPLG
(6)	CONTOUR AEROSPACE
(7)	COMPOSITE STRUCTURES DIVISION
(8)	HEXCEL STRUCTURES
(9)	SGL CARON COMPOSITES, HITCO
(10)	KAMAN AEROSPACE CORP.
(11)	LUCAS WESTERN, INC.
(12)	BOEING - STL
(13)	BOEING - CA
(14)	BOEING - MACON
(15)	MICHELIN AIRCRAFT TIRE CORP.
(16)	MISC SUB/EQUIPMENT
(17)	NORTHROP - GRUMMAN CORP.
(18)	NORTHWEST COMPOSITES
(19)	PRATT & WHITNEY
(20)	MARION COMPOSITES
(21)	REYNOLDS METALS
(22)	GKN WESTLAND AEROSPACE

(17) TRAILING EDGE
(17) VERTICAL STABILIZER TIP FAIRINGS
(17) ACCESS DOORS
(17) UPPER LEADING EDGE
(8) WING TO FUSELAGE FAIRING SKINS
(8) FILLET ACCESS DOORS
(11) PITCH TRIM ACTUATOR AND GEAR BOX
(17) PITCH TRIM FAIRING
(17) ELEVATORS
(17) HORIZONTAL STABILIZER
(17) HORIZONTAL STABILIZER LEADING EDGE
(17) HORIZONTAL STABILIZER BOX
(13) VOR ANTENNA COVER
(17) VERTICAL STABILIZER LEADING EDGES
(17) VERTICAL STABILIZER BOX ASSY
(4) DORSAL FIN AND ANTENNA
(8) WING TO FUSELAGE FILLET FRAME SUPPORTS
(12) CARGO DOOR
(14) UPPER AFT FUS CLOSE-OUT PANEL
(20) AFT RADOME
(9) TAIL CONE AND RUDDER FAIRING
(13) AFT CMPT DOOR
(13) AFT FUSELAGE
(13) CENTER FUSELAGE
(12) CARGO FLOOR ROLLERS/INSERTS
(13) CARGO FLOOR ASSY
(17) RUDDERS
(16) STRAKE
(16) RAMP ACTUATORS
(7) SPOILERS
(14) RAMP TOES
(12) CARGO RAMP DOOR
(20) FLAP TRAILING EDGE
(13) JUMP DOOR LEDGE
(1) TRAILING EDGE PANELS
(14) FLAP
(22) VANE
(16) INBO TE ISLAND
(14) AFT CLOSE-OUT PANEL
(14) JUMP DOOR
(14) FWD CLOSE-OUT PANEL
(13) UPPER ACCESS DOOR AND DITCHING ESCAPE
(13) FWD FUSELAGE
(12) MLG POD FRAME SUPPORTS
(1) OUTBO TE ISLAND
(17) AILERON
(12) TORQUE BOX ASSY
(14) JUMP DOOR
(17) UARRSI
(12) TROOP DOOR AIR DEFLECTOR
(13) TRAILING EDGE SUPPORTS AND FLAP, SPOILER AND AILERON HINGES
(6) STRINGERS
(21) WINGSKINS
(4) WING TIP ASSY
(10) RIBS AND BULKHEADS
(6) SPAR WEBS & CAPS
(12) FWD NOSE ASSY
(14) CREW ENTRANCE DOOR
(16) WINDSHIELD ASSY
(12) FWD PRESS BULKHEAD
(18) STRUT CMPT DOORS
(18) MLG POD SKINS
(20) WINGLET
(14) PYLON STUB ASSY
(12) ENGINE PYLON
(12) LWR FUS ACCESS DOORS
(16) RADAR
(20) FWD RADOME
(20) UPPER MAIN LANDING GEAR DOORS
(18) STRUT DOOR
(5) MAIN LANDING GEAR
(3) AIR-CONDITIONING PACK
(14) FIXED LEADING EDGES
(3) APU AND EXHAUST DUCTS
(18) AC ACCESS DOORS
(14) SLAT TRACKS
(14) SLATS (3 & 4)
(14) SLATS (1 & 2)
(5) NOSE LANDING GEAR
(15) NOSE TIRES
(2) NOSE WHEEL
(12) ACCESS DOOR
(20) NOSE LANDING GEAR DOORS
(2) MLG WHEELS/BRAKES
(15) MLG TIRES
(18) STRUT DOOR
(20) LOWER MIN LANDING GEAR DOORS
(13) AC RAM INTAKE AND EXHAUST
(18) DROP-OUT GENERATOR DOOR
(1) FLAP HINGE FAIRINGS
(14) PYLON FAIRINGS
(17) PRIMARY NOZZLE, DUCT, AND CORE THRUST REVERSER
(17) FAN THRUST REVERSERS
(17) FAN COWL DOORS
(17) INLET ASSY
(19) F117-PW-100 ENGINE
(17) FAN DUCT
(20) NLG STRUT DOORS
(14) EMERG EXIT DOOR

BOEING

This diagram from late 1998 gives a good perspective of the assembly breakdown of the C-17 as well as some of the suppliers contributing to the program. The suppliers have undergone some change over the years, and may yet see more, as names change and more manufacturing economy is sought. (Boeing)

A Humvee in the foreground and an M551 Sheridan light tank beyond are on pallets in the cargo compartment of the Globemaster III. They were later airdropped over Fort Bragg, North Carolina, on 27 June 1995 as part of the operational testing of the C-17. Although dropped in flight test, the Sheridan was later removed from service. (DoD)

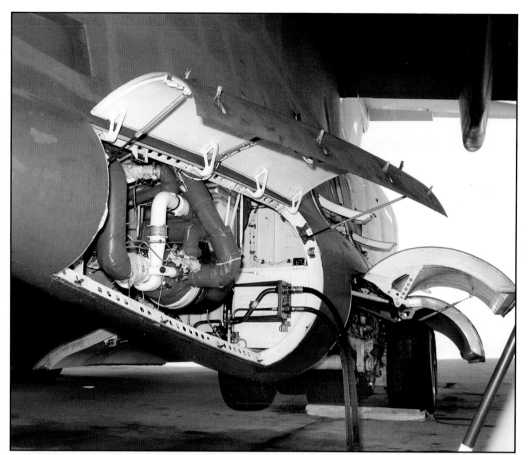

One of the two air-conditioning (AC) packs, this in the forward end of the port sponson, is exposed by the open maintenance panel. Hydraulics are being serviced and the ram air intake for the pack is open above the access panel. The other pack is in the forward portion of the starboard sponson, just aft of the APU exhaust port. (Boeing)

The sponsons are attached to tools that will precisely mate them up with the fuselage. The aircraft is being carefully positioned in preparation for this operation. Note the forward pressure bulkhead to which the radar will be attached later. (Boeing)

The size of the C-17's wing is evidenced by the author standing near the number 1 engine pylon. With the flaps lowered, the dual slots created by the aerodynamic vane and the slats are made clear. Outboard is the aileron and winglet. (Tony Landis)

Aircraft T-1 performs flutter testing with its cargo door and ramp open. Note the exciter vanes at the tips of the wings and horizontal tail. The tail cone has been removed and a High Attitude Recovery Parachute (HARP) system installed. (AFFTC)

The fan and core case of the F117 engines are readily accessible by opening the four cowling doors (two seen here) and propping them up with stay rods. The aft cowling doors incorporate the ducts for the fan airflow. (AFFTC)

The United Kingdom's Royal Air Force will begin receiving the first of four leased Globemaster IIIs in 2001. The aircraft will be based at RAF Brize Norton and may appear as shown in this retouched photo. (Boeing)

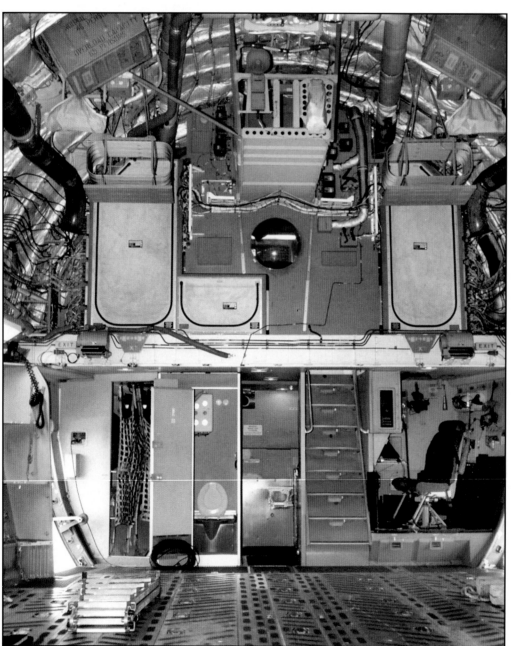

The forward end of the cargo compartment includes, from left to right, the crew entrance door, an equipment locker, the lavatory, the galley, the stairs to the flight deck, the loadmaster station, and an escape hatch (open). Above may be seen the crawlway to the centerline maintenance/ditching hatch, the two life raft containers under the FEDS hatches with coiled rope ladders, the window from the crew rest area, and access panels to avionics racks. (Bill Norton)

C-17A DESCRIPTION

A DREAM BECOMES A REALITY

The design of the C-17 was specifically aimed at ease of maintenance and easy operation by a minimal crew of just three individuals. This demanded great reliability, easy identification of faults so that a mission could continue and the problem quickly isolated and remedied later, and minimal need for the crew to monitor systems during the flight. The application of state-of-the-art systems made this possible, as will be detailed.

AIRFRAME

The C-17A structure is of conventional semi-monocoque construction possessing the high strength required for high-speed, low-altitude ingress missions at high gross weights, plus high sink rate STOL landings. The primary structure consists predominantly of aluminum with some steel and titanium, plus composites in secondary structure. The troop doors incorporate small windows and two additional windows are provided in the forward fuselage within the escape hatch and just aft of the crew entrance door. These allow scanning of the aircraft's exterior condition in flight.

The starboard forward fuselage has an emergency escape hatch. An overhead maintenance/ditching hatch provides access to the top of the aircraft via the crew rest area for use in the event of ditching at sea or for routine maintenance actions. This hatch is reached via a suspended crawlway accessed through a small door behind the top bunk in the crew rest area. Four additional overhead ditching Flotation Equipment Deployment System (FEDS) escape points are provided. When the FEDS is activated, explosive lines cut the aircraft skin at the four points to produce escape holes while automatically deploying 46-man life rafts. Personnel reach the holes via rope ladders that drop from the cargo compartment ceiling.

The cargo compartment floor is constructed of titanium and aluminum. Hatches, one at forward end of the cargo floor and the other just forward of the ramp on the bottom of the aircraft, allow access to a lighted tunnel beneath the floor. A manually operated creeper can carry a prone person along in this bilge area to conduct inspections or affect repairs.

The wings are three-spar structures with supercritical airfoil sections for reduced drag at transonic airspeeds and increased internal fuel volume. The requirement for combat maneu-

Aircraft P-19 rolls-out after a landing at Roosevelt Roads, Puerto Rico. For the deceleration, all the spoilers and thrust reversers have been deployed. The jet's finish appears to have faded and been touched up, with older markings painted-over. (DoD)

This photo of the left-hand ACM seat shows the vast number of circuit breakers on the face of the Electrical Power Center behind. The pilot side window has been slid back and a movable sunshield hangs down before it. Note the sliding transparent sun shade panel hanging over the side window. (AFFTC)

vering capability and longevity in a high-buffet flight environment has produced a wing that alone makes up nearly a third of the overall empty weight of the aircraft and which has a lower surface skin inches thick at the root. The upper and lower wing surfaces have elliptical plates to allow physical access to the internal fuel tanks. The composite winglets provide the benefits of greater wing aspect ratio without

the disadvantages of an actual increase in span. They also reduce tip vortex drag and vortex intensity for an overall drag reduction of as much as three percent. Stub engine pylon mounts, to which the individual engine pylons attach, are integral to the wing structure.

The vertical tail is equipped with twin, non-interconnected, two-segment rudders. The rudders' for-

ward and aft segments are articulated such that the aft segments deflect more than the forward segment for a more efficient airfoil. Horizontal gaps between the rudders and the stabilizer prevent interference during tail flexure. The horizontal stabilizer is equipped with twin, non-interconnected, two-segment elevators. A stabilizer trimming feature is provided by a jack screw elevator mechanism. A plug ceiling hatch above the cargo door, reached via a ladder, allows access to a tunnel with ladder rungs within the forward section of the vertical tail. This permits maintenance personnel to easily service equipment within the tail and to access the top of the horizontal stabilizer through outer hatches.

All of the control surfaces except the flaps and slats are made entirely of carbon composite. The flaps are a combination of titanium, aluminum, and composites, and the slats are aluminum with titanium heat shields. The C-17 possesses a total of 29 control surfaces, including horizontal tail trim. Each wing is equipped with four leading edge slat segments, an aileron, two flap segments, and four spoilers mounted just forward of the flaps. Each flap segment includes a forward span-wise aerodynamic vane that produces a double-slotted Fowler Flap configuration. A fixed slot is created between the vane and primary flap segment and the second slot is produced between the vane and the spoiler segments when the latter are stowed. The slats extend on rails and include extensible ducts to port heated engine air into internal slat cavities for deicing. All control surfaces are driven by individual pairs of hydro-mechanical actuators, excepting the spoilers which employ a single actuator each.

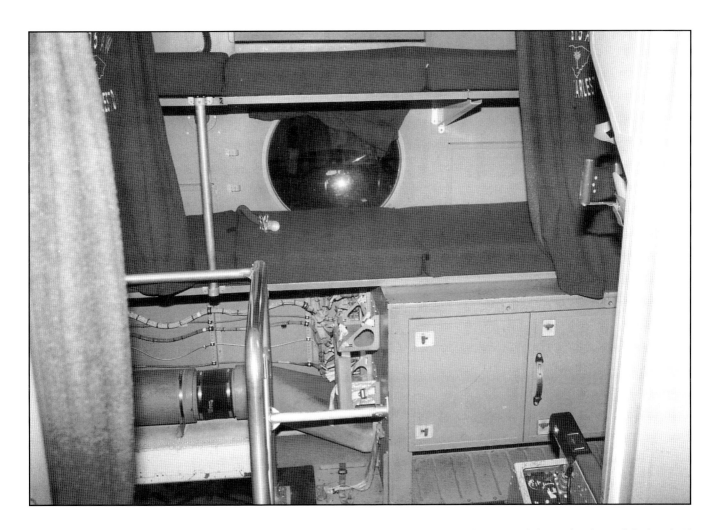

This image shows the major portion of the crew rest area with the two aft-facing seats (arm at right), two bunks, and the handrail beside the stairs to the cargo compartment. Note the window into the cargo compartment behind the bunks and the bottom of the panel (top of photo) covering the crawlway to the overhead maintenance/ditching hatch. (Bill Norton)

FLIGHT CONTROLS

The C-17 uses control sticks vice yokes to avoid obscuring the instruments or control interference from the pilots' legs. A center stick is used instead of the side stick controller because of U.S. Air Force preference and because the side stick would not provide the leverage required to fly the aircraft via the mechanical back-up control system (necessary for a tactical military airlifter). Only the upper portion of the stick column displaces laterally while the entire column displaces fore and aft.

The quad-redundant digital fly-by-wire EFCS and autopilot augments aircraft stability and provides specialized flight control modes to enhance mission effectiveness. Four independent flight control computers (FCC) manage the system. The FCC comparison and voting logic ensures error detection for system redundancy, with the system operating with as few as two of the computers online. This also provides a fail-operational feature with all flight control axes functional in critical modes even with two FCC failures and with failures passive such that no adverse inputs result. The EFCS replicates classical stability and control augmentation system functions, such as aileron/rudder interconnect and yaw damping, and also allows

relaxed static stability that permits a reduction in tail size. A pitch and roll axes autotrim feature relieves sustained trim commands to allow full utilization of axis control authority. Elevator and aileron trim are commanded via a quad-directional switch on top of each stick. Conventional rudder trim employs a knob on the center pedestal.

The EFCS possesses a multitude of flight control modes, tailoring aircraft response to particular flight regime requirements. Worthy of note is the Yaw Emergency Power Mode that moves only the lower rudder segment for operation on emergency battery power for 15 to 30 minutes.

This feature is necessary because of a predicted lightly damped or unstable dutch-roll oscillation with EFCS de-powered at high altitudes, requiring emergency yaw damping until the pilot can descend to an altitude at which the oscillation would become stable.

A three-axis mechanical reversion flight control system and manual pitch trim provides back-up in the event of partial or total EFCS failure.

This allows pilot stick motion to directly move ailerons, elevators, and lower rudder hydraulic actuator control valves via cables, pushrods, and so forth, rather than the electrical system performing this function. The system is "single-thread" with only one non-redundant control run for each axis providing an emergency "get home" capability. The overload and sensitivity protection in the yaw axis is provided by the upper rudder segment being automatically locked-out when the system is in mechanical control.

The flaps and slats are controlled together with the flap/slat handle on the center pedestal. This includes options for flaps up with slats extended, as well as the 1/2, 3/4, and full flap detent settings that include slat deployment. A flap setting greater than 1/2 automatically places the flight control system into a configuration to facilitate the backside control technique. A flap index switch sets the actual flap deflection angle. The index settings, 0 to 100, are dictated by performance variables for the particular flight condition, computed by the mission computers, and displayed on an MFD page. For example, a flap setting of 1/2 or 3/4 may produce the exact same flap position using different index settings.

Stall warning is a complex function of many aircraft and flight condition variables. When a stall is imminent the computers activate pilot stick shakers and provide an aural warning to the flight crew.

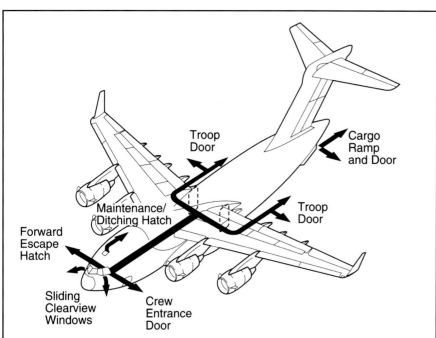

This diagram depicts the emergency ground egress points and locations for the C-17. The overhead maintenance and ditching hatch is accessed via a crawlway above the crew rest bunks. (Boeing)

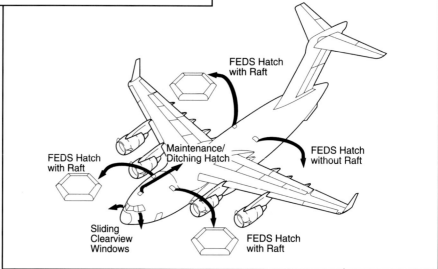

The emergency egress points in the event of water ditching are in the top of the aircraft since the rest of the jet will be underwater for the time it remains afloat. The aircraft is unusual in that four of the overhead exits used in the event of ditching are created by pyrotechnics cutting openings in the skin. This reduces weight and the maintenance actions associated with normal plug hatches. (Boeing)

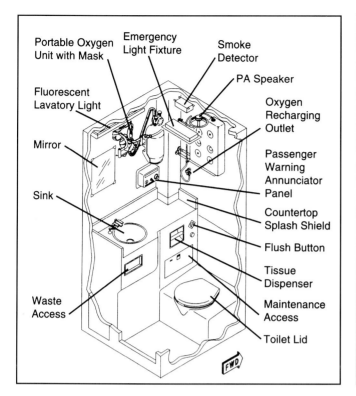

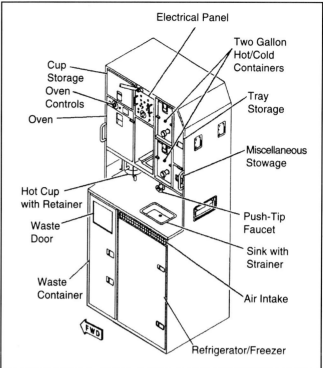

The lavatory on the C-17 is more like that found in an airliner than in other military transports. Its accouterments are detailed here. (Boeing)

Long-range missions on the C-17 make the galley a very welcome aspect of the C-17. Its features are identified in this drawing. (Boeing)

The Automatic Flight Control System consists of the automatic pilot, the autothrottle system (ATS), and the flight director (FD). The FD provides the pilot with the cues necessary to perform distinct tasks such as a manual instrument landing system approach and can automatically fly the aircraft in an autocoupled Category II instrument approach. The ATS adjusts the physical throttle setting for optimum cruise performance and to meet mission computer parameters.

AVIONICS

Voice communication is provided by two VHF, two UHF, and two HF radios. Several of these radios are provided with secure voice/data capabilities, anti-jamming features, plus satellite communication (SAT-COM) and UHF Line-of-Sight links. The laptop computer printer at the

forward loadmaster station is the data terminal for the non-voice communication. One Identification Friend or Foe transponder is also included. The UHF radios have automatic direction finding capabilities. The navigation radios consist of two VOR/ILS/Marker Beacon receivers, two distance measuring equipment receivers, and two GPS receivers. Thirty antennae on the fuselage are associated with these transceivers.

Primary control of all radios is performed by the integrated radio management system with aircrew interface through the communication-navigation control heads at each pilot station and the mission and communications keyboards in the center pedestal. These allow the programming and switching of all frequencies, among other functions, and eliminate individual radio control heads. There is a single backup radio

control head provided in the overhead panel for independent control of one UHF transceiver. Sixteen intercom control system panels are available throughout the aircraft. A wireless interphone system allows intercommunications in and about the aircraft without encumbering cords.

The weather radar provides weather avoidance, ground mapping, air-to-air skin painting, and beacon interrogation functions. Weather data may be displayed in color on the MFDs. The weather mode is effective to a range of 250 nautical miles and ground mapping to a range of 200 nautical miles. The radar can skin paint airborne targets and also track other aircraft transmitting X-band beacons, displaying range and bearing to each target.

A ground proximity warning system processes such data as the radar alti-

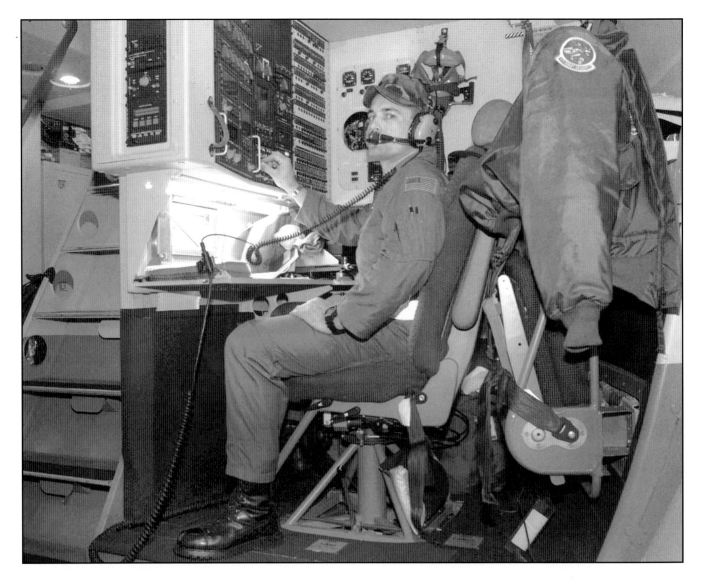

The loadmaster station of the C-17 is on the starboard side of the aircraft at the forward end of the cargo compartment. The controls at the station allow a loadmaster to perform airdrop functions without leaving the seat. Note the fold-down evaluator seat to the right. (DoD)

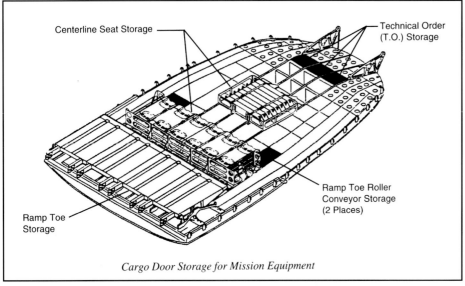

Centerline Seat Storage

Technical Order (T.O.) Storage

Ramp Toe Roller Conveyor Storage (2 Places)

Ramp Toe Storage

Cargo Door Storage for Mission Equipment

The ramp toes are stowed on the edge of the cargo door so they will not be in the way during operations where they are not required. They are easily shifted by one loadmaster to the end of the ramp while both the ramp and door are closed. Extra seats for the centerline of the cargo compartment, as well as litter stanchions, are also stored in the door as shown. (Boeing)

The interior of the cargo door, with the ramp closed up against its edge, is seen in this image. Visible are the ramp toes stowed on the end of the door and the cargo floor centerline seating rigs stowed within the door's structure. Note the large structural cross-members and long-stroke actuators. (Bill Norton)

Inspection of the structure under the floor of the aircraft is facilitated with a creeper running on rails along the centerline. This bilge space is accessed by a bottom hatch just ahead of the ramp hinge line and by an interior hatch at the forward end of the cargo compartment. Lying prone, crewmembers or maintenance personnel pull themselves along by hand. (Boeing)

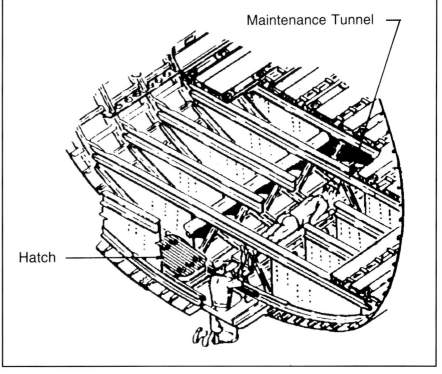

Maintenance Tunnel

Hatch

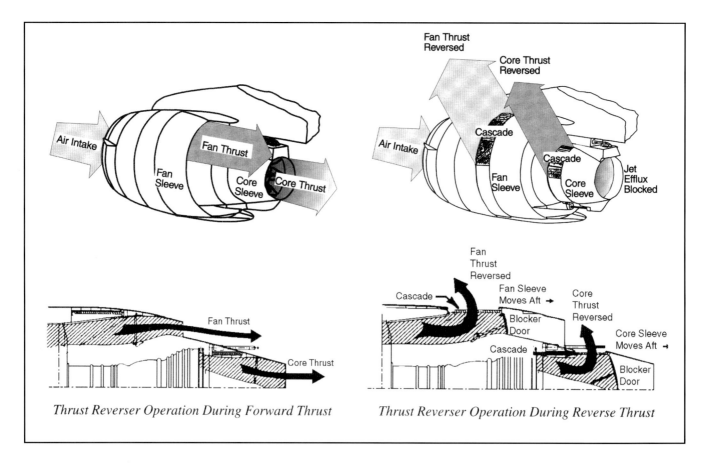

Thrust Reverser Operation During Forward Thrust

Thrust Reverser Operation During Reverse Thrust

Deployment of the F117 thrust reversers results in cowl sleeves moving aft to expose cascades while simultaneously positioning blocker doors in the fan and core exhaust paths. All engine exhaust is then redirected out of the cascades. These grills direct the air out and forward to help in aircraft deceleration. (Boeing)

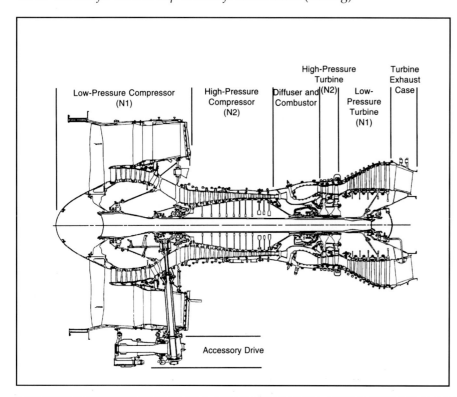

tude, vertical velocity, aircraft configuration, and glideslope deviation to alert the aircrew to the potential for an inadvertent collision with the ground or the incorrect configuration during an approach.

The extensive systems aboard the aircraft are managed by many individual computers, too numerous and complex to detail separately. Transfer and integration of all of the information from the various computers and systems are facilitated by a number of data buses.

POWERPLANT

The Pratt & Whitney F117-PW-100 engines are twin-spool, single stage fan, high bypass (5.9:1 ratio) turbo-

The individual sections of the F117-PW-100 high bypass ratio turbofan engine are labeled in this cutaway. (Boeing)

fans, producing 40,400 pounds of installed thrust. The engine can efficiently burn all standard turbine engine fuels without difficulty, enhancing operational flexibility. The engine includes a low-speed rotor of one fan stage, four compressor stages, and five low-pressure turbine stages. The high-speed rotor consists of twelve high-pressure compressor stages and two high-pressure turbine stages. The 6th and 10th stage rotor vane angles may be adjusted for improved starting and to prevent compressor surge. The combustion section consists of a single annular burner with 24 fuel injectors. All bleed air is derived from the 10th, 14th, and 17th stages. The engine nacelles have fixed geometry inlets that incorporate hot bleed air anti-icing ducts.

All control of the engines are electrical, there being no mechanical connection between the throttles and the engines, with operation regulated and optimized by a dual channel Electronic Engine Control (EEC) on each engine. Engine indications are provided on the MFDs, and the standby engine display (SED) supplements the MFDs and provides basic indications when the aircraft is operating on battery power alone. The EEC normal mode of control is engine pressure ratio (EPR). The actual EPRs resulting from throttle movement are determined electronically as a function of SED mode selection and appropriate operating limits. The SED modes include: Maximum for takeoff (five minute limit); Intermediate for additional thrust during continuous operations (such as engine-out climb); Maximum Continuous for normal climb; Derated Takeoff or the minimum thrust for a safe takeoff; and Manual in which the aircrew sets the desired EPR on the SED. Selecting any of these settings

will change the actual thrust produced throughout the throttle range as dictated by that setting.

The basic accessories driven by the engine accessory gearbox consist of two hydraulic pumps, a ground and flight pneumatic starter, a generator, an alternator to power the EEC, an oil pump, a fuel control unit and pump, and an oil scavenge pump. The pneumatic source for starting may be derived from an external cart, the APU, or another engine. Windmilling airstarts are also possible. Detection of an engine fire sounds an aural warning identifying the engine on fire and illuminates the fire handle to be pulled to discharge the extinguishing agent. Two discharges are available per wing, and can be used for either engine on that wing.

LANDING GEAR

The main gear retracts into sponsons mounted entirely external to the fuselage cavity. Landing loads are reacted by an air/oil shock strut mounted between the rear of the main vertical gear post and the pivoted truck. With the aircraft in flight and the gear extended, the shock strut is fully extended and the trucks pivot to a vertical position, "heel down," positioning the inboard wheels slightly above the center and outboard wheels.

On landing, the trucks rotate to a horizontal position as the strut compresses. During retraction the gear on either side of the aircraft rotates 90 degrees toward each other while lifting up into the sponson. There is a single extend/retract actuator and uplock latch for each gear, with smaller actuators providing the force to move an over-center uplock and downlock link. Gear uplocks may be manually released to allow emergency free-fall using T-handles in the cargo compartment adjacent to each gear well. Small windows into the wells allow for visual verification of gear position. The windows may be opened to permit installing

This photo shows the engine thrust reverser translating sleeves and cascade grills to good advantage on engine No. 1 of T-1. (AFFTC)

The underside of the flaps shows the unpainted titanium panel directly aft of the engine exhaust and the large hinge fairings. The flaps on each side of the aircraft are actually two separate panels but all move as a unit. Visible are the brackets attaching the aerodynamic vane to the primary flap structure. (AFFTC)

safety pins in flight. For rapid tire changes the main gear incorporates an integral jacking feature whereby the truck with the flat tire is raised off the ground by hyper-extending the shock strut on the adjacent gear.

The electronic anti-skid system automatically avoids blown tires on touchdown, in the event the brakes are being applied, and also stops tire rotation upon retraction. Separate emergency brake pressure accumulators allow up to 10 full brake applications in the event of hydraulic system failure as well as provide a parking brake feature. Two parking brake handles are provided on the cockpit center pedestal, one for each side of the aircraft. The accumulators may be manually pumped to full pressure if necessary using a handle in the cargo compartment. Fuse plugs in each wheel melt to relieve tire pressure and avoid explosive rupture should the tire become excessively heated due to hot brakes or high-energy taxi operations. Brake temperatures are available to the pilot on an MFD page and warning annunciations are provided to help ensure against overheat.

The main gear doors consist of forward and aft outer doors plus forward and aft inner doors. The outer doors rotate outboard sufficient for ground clearance and the inner doors open flush with the belly of the aircraft. Links to move the doors are mechanically connected to the gear and so are operated by the extend/retract cycle. This also eliminates the requirement for separate

With the nose gear extended only the small aft doors are open, as shown. The forward doors open only during the retract and extend cycle. The gear rotates forward during retraction. (AFFTC)

door latches or persistent hydraulic pressure to keep them closed. The outer doors may be released from their links and swung up and latched to the outside of the sponson to provide easier maintenance access. Upon gear retraction a large sponson panel (the upper outer door) rotates outward to allow for passage of the upper portion of the gear post beyond the profile of the sponson. With the gear extended, a much smaller panel (the upper inner door), at the center of the upper outer door, is opened to allow the top of the post to project beyond the sponson profile.

The single nose gear assembly retracts forward. It includes a conventional cantilever air/oil shock strut. Two small aft nose gear doors are open whenever the nose gear is extended to clear the strut. Two large forward doors are open only during gear transit unless opened for maintenance purposes. Dual extend/retract actuators and a smaller uplock/downlock link actuator drive the entire cycle.

Standard nose gear towing is available as well as the option of main gear towing. Towing via the nose gear with turns beyond the 65-degree limit requires that the strut torque links be disconnected.

PRIMARY SYSTEMS

The instrument panel MFDs are supplemented by the more familiar standby analog instruments in the event of MFD or electrical system failures. The "dark cockpit" concept allows indicators to remain dark provided the related system is performing nominally. Indicator lights can be switched on by indivual selection, with the exception of system flowlines. The automatic self-

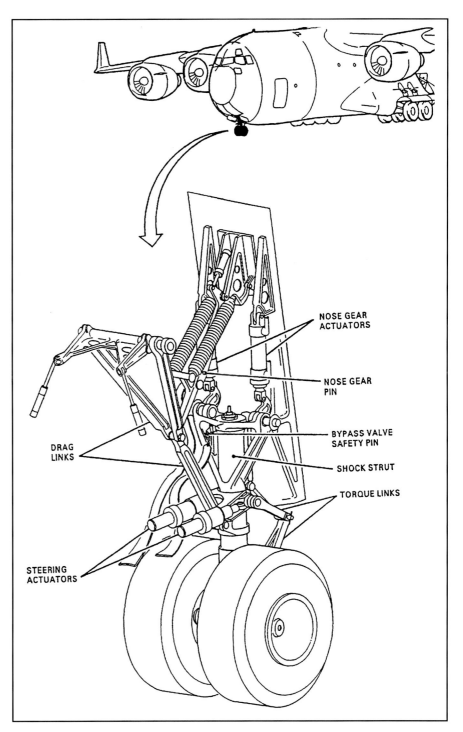

This drawing illustrates the nose gear arrangement. Apart from the two extend-retract actuators noted, a third smaller actuator at the top of the assembly serves to move the over-center uplock/downlock to permit gear movement. (McDonnell Douglas)

monitoring and warning systems alert the aircrew to abnormal system functions so that system control panels and status displays need not be routinely monitored. All this reduces pilot workload and pressing the illu-

minated switch light is often the only action required following a malfunction. The automatic warning systems include the standard master caution and warning lights, plus the warning and caution annunciation

This photo was taken with the outer gear doors raised to permit easier maintenance access. Note how the inboard wheel is placed slightly aft of the other two. The inner pair of doors, opened against the bottom of the fuselage, are visible along with some of the door links. (Tony Landis)

panel which provides a scrollable display of color-coded messages.

All of the fuel is contained in the wings in a "wet wing" configuration (no fuel bladders). Each wing contains an inboard and outboard fuel tank. The inboard tank has a 24,400-pound capacity aft portion and a 27,255-pound capacity forward portion. The forward portion contains a feedbox to supply the inboard engine. The 37,760-pound capacity outboard tank normally feeds the outboard engine via a feedbox. A vent box is installed at the wing tip to allow the air pressure within the tanks to equalize with the ambient pressure and avoid tank damage during altitude changes. A fuel dump mast exits the trailing edge of the wing between the aileron and flap. An onboard inert gas generating system (OBIGGS) is used to produce non-combustible gas from the

This shot shows the full length of a gear post with some of the many braces, links, and actuators. It is attached to the aircraft via very stout fuselage frames at left. The main gear is exceptionally rugged to support the STOL mission. (Tony Landis)

The top of the vertical tail is accessed via a hatch reached by a ladder positioned on the cargo door. Rungs within the tail allow personnel to climb up to inspect equipment within the tail or to climb out onto the horizontal stabilizer through a door at the top of the vertical. (Boeing)

Below: *The two main landing gears on each side of the aircraft are complex structures. The three-wheeled trucks rotate in opposite directions during retraction, aligning the wheels with the fuselage. The four gear doors per side are attached to the gear via mechanical links, opening and closing without individual actuators. (Boeing)*

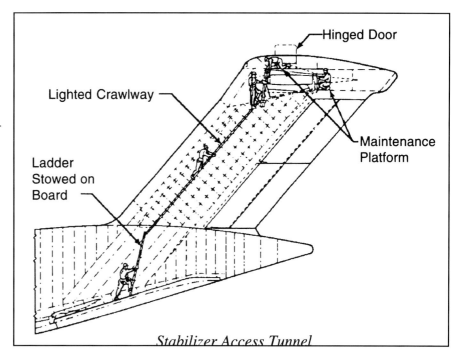

Hinged Door

Lighted Crawlway

Maintenance Platform

Ladder Stowed on Board

Stabilizer Access Tunnel

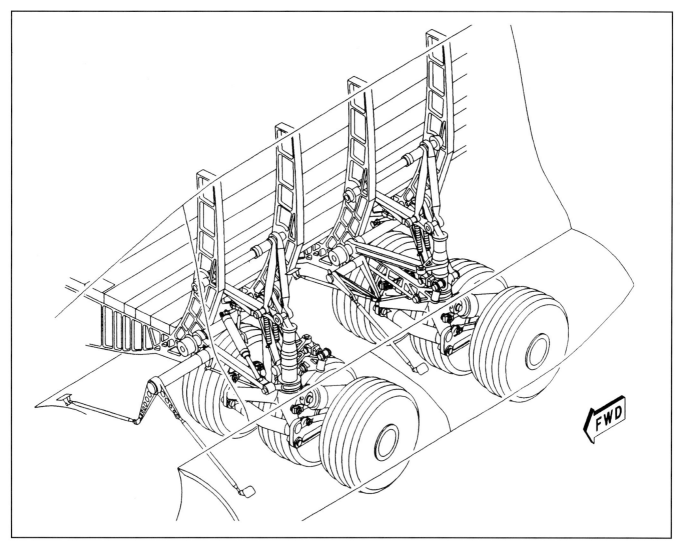

FWD

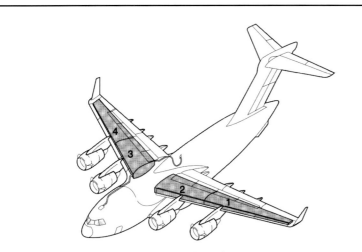

Fuel Tank Configuration

Fuel Tank	Maximum Usable Fuel			
	Pounds*	Kilograms*	Gallons	Liters
Left Outboard No. 1	37,760	17,128	5,636	21,335
Left Inboard No. 2	52,640	23,877	7,857	29,742
Right Inboard No. 3	52,640	23,877	7,857	29,742
Right Outboard No. 4	37,760	17,128	5,636	21,335
Engine Feed Lines	254	115	38	143
Total	181,054	82,125	27,024	102,297

* Based on JP-8 or Jet A-1 at 6.7 Pounds per U.S. Gallon (0.80 kg/liter)

This drawing shows the fundamental fuel tank arrangement of the C-17. The main points of fuel onload and offload are shown by the lines running to the aerial refueling receptacle and the single point refueling (SPR) station at the aft end of the starboard sponson. Not shown are the myriad of fuel lines, valves, and pumps. (Boeing)

inboard engines' bleed air to pressurize the vapor space in the tanks. Each tank contains two submersible boost pumps and the feedboxes contain a transfer pump. A myriad of valves and manifolds allows fuel transfer between any of the tanks and any tank can feed any of the engines. Should all of the tank pumps become inoperative, the engines can still be supplied via gravity feed and suction feed from the engine-driven pumps.

An automatic fuel management system pumps fuel between tanks as necessary to maintain an equal distribution between the two wings and to help keep the aircraft's CG forward. All of these functions may be performed manually and individual quantities monitored using the fuel system panel in the cockpit.

The aircraft has four completely independent 4,000-psi hydraulic systems. Design philosophy ensures flight control redundancy by having each of the two actuators driving any single control surface (except spoiler panels) powered by different hydraulic systems. A surface is func-

Ground refueling is performed at the rear of the starboard sponson. One of two fuel hoses are attached to fittings and the aircraft fuel system is operated from the panel just aft, as shown. Also seen to good advantage are the top of the gear posts protruding beyond the mold line of the sponson. (DoD)

tional on a single actuator, albeit with reduced capability. All this implies that the aircraft may theoretically fly on a single hydraulic system in an emergency. Four large hydraulic fluid reservoirs are mounted on the cargo compartment walls so that fluid levels may be visually checked during preflight.

The hydraulic systems are automatically regulated by one of two system controller computers, although manual operation is possible. A primary and a secondary engine-driven pump power each system. An electrical auxiliary pump provides hydraulic pressure when the engines are not operating. Additionally, the No. 4 system can be powered at a reduced capacity by a ram air turbine (RAT) stowed in

the forward bottom of the starboard gear sponson. The RAT may be extended in the event of complete engine failure. Once clear of the aircraft, the airstream turns the 27-inch diameter turbine.

Primary aircraft AC electrical power is supplied by four 75-90 KVA 3 phase, 115/200V, 400 cycle Integrated Drive Generators. Two hundred amp transformer rectifiers convert the AC power to 28 VDC. With the engines shut down on the ground, power may be derived from either the 90 KVA APU generator or an external power cart. Two 40-amp nickel cadmium batteries also supply AC/DC power for APU starting, ground refueling, and emergency electrical power for up to 30 min-

utes. Power is distributed via a multitude of AC and DC primary, transfer, emergency, and tie buses and relays to the various systems and avionics. More than 2,000 individual circuit breakers are distributed throughout the flight deck and cargo area. Over 125 miles of wire makes up the "nervous system" of the Globemaster III.

SECONDARY SYSTEMS

Four air data computers (ADC) perform all computations necessary to provide such data as airspeed, altitude, outside air temperature, and so forth, to the pilot displays as well as to such computers as the FCCs. Four pitot-static dogleg probes, two on either side of the forward fuselage,

Shot looking aft along the starboard sponson, the Auxiliary Power Unit (APU) is exposed. The APU is an Allied-Signal small gas turbine engine used as a self-contained electrical power source during ground operations and as a means of starting the aircraft's engines. (Tony Landis)

This photo shows the "porch" and deflector outside the paratroop door that makes jumping from the C-17 a bit easier than from other airlifters. Note the retracted aft corner of the fairing, reducing a trip hazard. The bright panel just above the porch at the forward end is a light. (Bill Norton)

provide the air pressures to the ADCs. Six AOA vanes, three on either side of the nose, and two total air temperature probes on the bottom of the nose also provide air data.

The environmental control system (ECS) functions include air conditioning, avionics cooling, cabin heating, aircraft pressurization, ice protection, smoke detection, windshield defog and anti-ice, and the windshield wipers. Air for most environmental requirements are supplied by engine bleed, although APU and ground cart air is also available on the ground. An air conditioner pack is installed in the forward end of each gear pod. Ram air may be used as a backup air source and is drawn in through an inlet at the forward end of each pod. Engine bleed air is available for mixing with pack air to achieve the desired cabin temperature. Cargo compartment heating is

supplemented by heating elements below the floor.

The fuselage can be pressurized by engine bleed air to maintain a cabin pressure altitude of 10,000 feet or below and is automatically regulated as the aircraft climbs and descends. Normal pressure venting is accomplished through a square valve aft of the nose gear. Additional pressure venting or emergency depressurization is accomplished through three round valves on the starboard side of the fuselage above the cargo ramp. Automatic negative pressure relief flapper valves are also installed in the aft fuselage (two elliptical ports on the port side of the fuselage above the cargo door and one on the starboard side).

A liquid oxygen converter and associated equipment supply crew oxygen, distributing oxygen to regula-

tors throughout the aircraft. A number of portable oxygen bottles with refill lines and quick-don masks are also provided. Likewise, passenger and auxiliary oxygen converters are installed. Airline-type oxygen masks are provided for passengers and additional oxygen connectors are available at each of the 54 sidewall seats to supply oxygen for HALO paratroopers.

Potable water is available and the aft cargo compartment contains two urinals. First aid kits, hand fire extinguishers, crash axes, and other emergency equipment is available. A series of 22 photosensitive smoke detectors are located throughout the aircraft with panels indicating the location of the detector sensing smoke. Additionally, crew stations are provided with receptacles for chemical warfare suit ensemble power.

Six foldout jack attachment points, a pair on both sides of the forward fuselage and two pair on both sides of the aft fuselage, are provided for raising the aircraft off of its landing gear. This shot shows jacks positioned under the two most forward of the attachment points. (Tony Landis)

PRODUCTION 5 & OPS

SLOW BEGINNINGS AND A NEW STANDARD

The acquisition of a new military aircraft is a very costly undertaking and one that a nation's leadership must consider very carefully. If the system fails to live up to expectations it may adversely impact the country's ability to defend itself or meet its national objectives. The C-17 was no different from all such U.S. efforts, with controversy surrounding it for many years until the jet was ready to demonstrate its value. When this hurdle was overcome and production came up to full rate, the Globemaster III quickly established its unique place among the U.S. Air Force's stable of airlifters.

WHITHER THE C-17?

Throughout most of flight test and the first few years of production, the future of the C-17 was being constantly challenged in the nation's capitol and its continuation was by no means certain. The end of the Cold War and the subsequent cuts in defense spending prompted, in April 1990, a reduction in the number of aircraft to be purchased from 210 to just 120 and further cuts were constantly threatening. Matters were not helped by the lengthening schedule and the high cost of the aircraft (the most expensive transport aircraft ever developed and the third most expensive aircraft in history).

The system deficiencies typically uncovered in flight tests were widely reported in the press and fed doubts about the suitability of the aircraft. The dispute resulted in a reduction in the initial annual buys of jets with no aircraft bought in 1991 (although there turned out to be no actual pause in production). The Secretary of Defense announced in December 1993 that there would be no commitment beyond 40 jets until the program could demonstrate over the next two years that it could produce affordable aircraft that met mission requirements.

In 1993 Congress directed that the Department of Defense (DoD) study alternatives to a full buy of C-17s. This came to focus on a possible mix of Globemaster IIIs and a Non-Developmental Airlift Aircraft (NDAA) identified as the C-33. Likely NDAA candidates included the Boeing 747-400F, the McDonnell Douglas MD-11F, and a notional Lockheed C-5D with new engines and a glass cockpit. Such aircraft would have had superior range and payload capabilities when compared with the C-17 but would have lacked the flexibility and tactical utility. The great success of the RM&AE and MDA's strides in reducing program cost gradually won the C-17 a new measure of acceptability. This was instrumental in the decision on 3 November 1995 to acquire the full 120-ship fleet at an accelerated rate.

The buy of the next 80 aircraft remained to be negotiated. Unlike previous major weapon systems, the DoD won approval to commit to a seven-year $16.2 billion purchase of the full 80 jets. This, the largest and longest multi-year defense contract ever awarded, was signed on 31 May 1996. This provided as much as $1.025 billion in cost savings and guaranteed a unit price of the aircraft from $183 to 172 million across the years of manufacture. At a typical production rate of 15 aircraft per

Aircraft P-10 flies low off the South Carolina coast. Unlike earlier high-capacity military transports, the C-17 was specifically designed to withstand the buffeting experienced during low altitude flight. (DoD)

The huge wing box structure is assembled in a jig, in the orientation seen here, by automated machinery. (McDonnell Douglas)

year, the last of the 120 aircraft will be delivered in 2004. However, DoD budget pressures in 2000 saw three aircraft to be bought in 2001 deferred to 2003. The entire C-17 program will total about $43 billion.

PRODUCTION

The C-17s are built in Boeing's giant Building 54 at the Long Beach Airport with the words "Home of the USAF C-17" emblazoned on the front. Production of a C-17 requires the teamwork of over 1,000 suppliers and 15,000 people. The prototype aircraft required approximately 2 million man-hours to construct and this has been progressively reduced to less than 750,000 hours on later production examples. Boeing had made changes to the assembly line and to some aircraft components to simplify construction and reduce overall costs. Between aircraft P-9 and P-17 the time needed to build a C-17 had dropped from 505 to 389 days. Boeing's success on the production line since P-12 was measured by successive deliveries of aircraft ahead of schedule.

The company's hard work on airlift and tanker programs was recognized with the 1998 annual Malcolm Baldridge National Quality Award for Manufacturing. The efforts went a long way in winning the confidence of the U.S. government and potentially bringing further orders.

The center fuselage structure is assembled inverted and then moved to the next station in the assembly line before being righted. (McDonnell Douglas)

REVISIONS AND UPGRADES

As is typical, changes were required to the early production aircraft to correct system deficiencies found in flight tests and initial service. Much of this retrofit work was performed in Tulsa, Oklahoma, while production changes were introduced into the Long Beach production line. These included electrical system upgrades, OBIGGS and fuel system changes, modifications to the wings and fuselage to bring them to full strength, changes in flap and slat material to increase their tolerance to high engine exhaust and reverser temperatures, and more minor upgrades.

As production got up a head of steam, further changes were introduced into the new aircraft, all of which were identified with block number designations coincident with the annual production lot numbers. Many of these changes are retrofitted at the earliest opportunity to the machines already delivered, either in the field or at depot. Many were computer upgrades, software revisions, and other small changes intended to resolve minor problems or to take advantage of new technologies. Among the latter was replacement of the three mission computers with two powerful core integrated processors. Some were easily implemented improvements but some represented more substantial changes like redesign of the APU exhaust ducting to reduce airframe heating and hot gas ingestion by the No. 3 engine.

McDonnell Douglas Aerospace made changes to the assembly line and to some aircraft components to simplify construction, reduce weight, and cut overall costs. Among these were revised landing gear pods and the Nacelle/Engine Affordability Team (N/EAT) nacelles that improved construction efficiency. In the same vein, Boeing redesigned the C-17's horizontal tail, replacing some components with composites. Although the center box beam of the tail remained

A busy production line shows aircraft in various stages of construction. In the foreground both wings have been mated to the fuselage while farther along the line the primary box structure of the vertical tail has been attached to the aft fuselage of another airplane. (McDonnell Douglas)

The Building 54 assembly facility at Long Beach Airport, California, was expanded to 1.1 million square feet to accommodate the C-17 production. Now emblazoned with the Boeing logo, when photographed two brand new C-17s were in final preparation for delivery to the U.S. Air Force. (McDonnell Douglas)

Pilots spend considerable time in the C-17 motion-based simulator cab during initial and proficiency training. This and computer-based training helps to reduce the cost of operating the C-17 fleet. (Boeing)

metallic, the use of composites for the outer torque box and other elements reduced the tail's weight by 20 percent and greatly simplified assembly for a substantial reduction in recurring cost. The first revised tail was flight tested on T-1 in March 1999 and then installed on P-51 (the first of the Block 10 configuration) for delivery. There are no plans to retrofit the tail on earlier aircraft.

Some engine improvements were introduced as the '94 Package with aircraft P-20. These helped to reduce operating temperatures and noise, while improvements in the EEC increased efficiency for about one percent gain in specific fuel consumption. The MDA efforts at improving cruise performance by reducing aircraft drag, with such measures as improving the seals around control surfaces and revising the flap trailing edge, were introduced on P-30. The SKE 2000 system has also been added to the aircraft, allowing the same capabilities as the baseline SKE with lighter and more reliable equipment.

A major revision of the OBIGGS is also planned in the future and improvements to the pressurization system are in work. Avionics improvements are also planned to permit the C-17 to operate within the global air traffic management systems introduced in 2000. This will see the substitution of new radios plus ring laser gyro inertial navigation system (INS) units with integral GPS receivers for accurate position monitoring capabilities. A Traffic Collision Avoidance System and new weather radar with predictive wind shear capability are to be installed at some future date.

Other aircraft changes have added to the operational utility of the air-

craft, some in response to operator needs and others to improve the performance of the jet. The original capacity of aeromedical litter stanchions for 48 patients was reduced to 36 as revised litter spacing requirements for all U.S. Air Force aircraft were introduced that allow easier access to patients.

New airdrop configurations were cleared permitting the dual row gravity drop of equipment pallets and drops off the ramp. The new capability employed up to 12 463L pallets or 8 of a new 16-foot Type X pallet for long items like "Humvee" vehicles and howitzers weighing up to 10,000 pounds. These capabilities reduce the time and number of aircraft (48 rather than 68) required to put a U.S. Army brigade on the ground with all of its gear. Testing was also performed to clear the aircraft for 20-second interval takeoffs and to permit closer in-train formations for the brigade airdrop. All this effectively answered an outstanding issue about the jet's capabilities to take over the C-141B's tactical airdrop missions and allowed 32,000-foot aircraft spacing during the airdrop. It also had the effect of increasing the C-17 fleet's airdrop capacity by 266 percent. The first example with this capability was delivered in 1999 with aircraft P-51 and earlier aircraft were then retrofitted.

Another major change adding considerable capability to the Globemaster III is the extension of the inboard fuel cells into the centerwing dry bay, adding about 10,000 gallons (67,000 pounds) of capacity. This will add approximately 1,000 nautical miles of range with a maximum payload. It will enhance many missions but especially permit deployments more quickly and economically by reducing the number

A pair of maintainers monitors the preflight checks of a C-17 at Ramstein AB, Germany, on 14 April 1999. Water from the apron is sucked up into the number 4 engine by the inlet vortex. The Charleston AFB jet is preparing to depart for Albania in support of NATO's Operation Allied Force. (DoD)

of aerial refuelings or en route stops. The extended range capability is to be introduced beginning with P-71 and eventually retrofitted to all earlier aircraft. Boeing is also exploring the potential of expanding the maximum gross weight of the aircraft to 600,000 pounds or more.

During the winter of 1995–96, in Operation Joint Endeavor, there was concern about small arms fire during flights into Tuzla and Sarajevo. Ceramic/Kevlar composite armor was installed with Velcro-type material to the cockpit floor, bulkheads, and seat bottoms. This gear was

Aircraft P-8 at Howard AB, Panama, on 14 December 1994. Note the landing/taxi light on the forward fuselage and the scanning light on the sponson. (DoD)

developed and deployed to the field in just two weeks. Joint Endeavor also saw the first operational use of the interim Airlift Defensive System for countering IR-guided antiaircraft missiles. The equipment consisted of passive infrared sensors at the four quadrants of the aircraft and either automatic or manual deployment of flares from dispensers mounted within the airframe. Manual operation is via a control head mounted aft of the throttles. The system was first tested on aircraft P-7 at Eglin AFB, Florida, during the summer of 1994. This entire effort, from start of design to deployment, was executed in just four months.

C-17 UNITS

Following the decision to buy a fleet of 120 machines, the fundamental plans were to convert two active-duty U.S. Air Force wings, each with 48 C-17s, along with an Air National Guard (ANG) squadron and a small training squadron. This fleet distribution provides 10 backup aircraft to permit full-strength operations while others are away at routine depot maintenance visits.

When the full production decision was made the first squadron of jets was already operational. Since 14 June 1993, with aircraft P-6, aircraft were delivered directly to the 17th Airlift Squadron (AS) of the 437th Airlift Wing (AW) at Charleston AFB, South Carolina. On 17 January 1995 the 17th AS was declared operational with 12 aircraft, 48 crews, and a fully trained cadre of maintenance personnel. The 437th AW continued receiving C-17s to convert two other squadrons from C-141Bs. The second to change over was the 14th AS (named the "Pelicans") receiving its first C-17 on 18 February 1995 with aircraft moved from the 17th AS in what became a common procedure. The 15th AS "Global Eagles" was stood-up on 22 September 1997. In tandem with the 437th AW's conversion its U.S. Air Force Reserve partner, the 315th AW (with the 300th, 301st, and 317th AS), also at Charleston AFB, converted to the Globemaster III. These units are actually associate squadrons, sharing the same aircraft as their counterpart active operators.

During December 1996 the C-17 aircrew training activities were moved from Charleston AFB to the 58th AS, 97th Air Mobility Wing (AMW), at Altus AFB, Oklahoma. The first of the 97th AMW's aircraft (P-18) arrived on 23 March 1996 and the last on 10 November 1997 with a

An impressive load-out includes three OH-58 Kiowa scout and two AH-64 Apache attack helicopters. The rotor blades of the Apaches have been removed but the outboard Kiowas appear to have been loaded without apparent disassembly. (AFFTC)

total of eight airframes. The 58th AS will train some 800 Globemaster III aircrew per year when all C-17 units are fully manned. It was probably not surprising that the 97th AMW's instructors beat other C-17 teams during the Globemaster III's first competition at the annual Airlift Rodeo in 1996.

The 62nd AW at McChord AFB, Washington, began replacing its C-141Bs with the new airlifter on 30 July 1999 with the delivery of P-51 and P-52. The first of the 62nd AW's units to convert was the 7th AS, followed by the 4th and 8th "Soaring Stallions" Airlift Squadrons. The 4th AS will actually be reactivated in 2003 after retiring its C-141s. The 62nd AW's associated Reserve squadrons of the 446th AW at McChord AFB were converting in tandem.

The 183rd AS, 172nd AW of the Mississippi ANG at Allen C. Thompson Field in Jackson, will also replace its C-141s with Globemaster IIIs, taking on six machines in July 2004.

Future-year defense budget planning includes 14 additional C-17s. These and another Globemaster III are to replace C-141Bs serving with the 16th AS "Bad to the Bone," 437th AW, with the SOLL II (Special Operations Low Level) mission. However, the actual aircraft configured for the mission may be drawn from among the initial 120 aircraft. It is planned that the modified aircraft would be available in 2004, reactivating the 16th AS after closing in 2000. The mission consists of long-range rapid-deployment of special operations troops at low altitude with night airdrop or insertion into blacked-out austere airfields. The SOLL II C-141Bs feature a forward-looking infrared turret under the nose,

A belly full of 82nd Airborne Division paratroops settle in for a short flight prior to a jump into Sicily drop zone, Ft. Bragg, North Carolina. The jump, along with 102 paratroops from another C-17, was performed on 27 June 1995 as part of the C-17's operational testing. (DoD)

SATCOM capability, expanded electronic countermeasures equipment, expanded flare/chaff dispenser capacity, and the optional use of a Plexiglas dome in place of the overhead maintenance/ditching hatch, the latter to permit full azimuth scanning for fighter opposition. The C-17 SOLL II configuration had not been defined at time of writing.

The United Kingdom's Royal Air Force (RAF) will begin receiving four Globemaster IIIs in mid-2001 on a seven-year lease, with two one-year extension options. The aircraft will be based at RAF Brize Norton with a squadron unidentified at time of writing.

OPERATIONS

Although the C-17 continued to suffer from "childhood diseases," with the landing gear providing the most maintenance challenges, they became fewer and did not prevent the jet from doing yeoman service

from its first days on line. As soon as the C-17 joined the 17th AS it began flying training sorties and the usual freight missions in CONUS. Even before the squadron was declared fully operational it had flown its first "real-world" emergency airlift missions, these to Kuwait during Operation Vigilant Warrior in response to Iraqi saber rattling. Beginning on 14 October 1994 the C-17 contributed two direct flights of 14 to 15 hours duration each. The first C-17 operational mission was flown on an aircraft delivered just a month before which had only 58 hours on its log. The crew picked up five U.S. Army vehicles, containerized cargo, and other supplies at Langley AFB, Virginia, before departing for the Middle East. The following summer, during the August 1995 Operation Intrinsic Action, C-17s again flew to Kuwait with the airlifter's participation in its first large-scale exercise.

Just six months after passing the RM&AE, the still-young C-17 force

The sky behind a C-17 fills with giant parachutes as it delivers a load of heavy equipment and other cargo using the new dual-row airdrop system. The capability proved a significant enhancement to the aircraft after it had already entered service. (AFFTC)

composite unit of heavy airlifters, dubbed Charlie Squadron, operating from Rhein-Main. The Globemaster contingent had originally consisted of only six aircraft until the U.S. Army saw what the aircraft could do and requested more of the airlifters to be brought over.

Combined with other NATO mobility assets, the jets participated in the largest military deployment in Europe since the Berlin Airlift of 1948-49. The mission was mostly high capacity, short-distance intratheater airlift of personnel and materiel. The first C-17 into the war zone landed at Sarajevo on 8 December with 154,000 pounds of humanitarian relief supplies. Because of small arms and antiaircraft missile threats the C-17 carried an extra crewmember to serve as a scanner and they flew many of their missions at night. The operation began in earnest in mid-December with the aircraft operating around the clock in very poor weather. The C-17 was critical to the movement of outsized and heavy cargo because of the confined surface area at the austere Tuzla Air Base, the Americans' forward deployment center. The jet's systems allowed it to spend an average of only 30 minutes on the ground at the destination, the same as a C-130 which brought in only a quarter of the Globemaster's capacity. This capability helped to keep up the fast pace and cargo throughput.

With the roads into the war zone badly congested, the C-17 allowed priority deployment of a mechanized infantry battalion, flying troops and 29 armored vehicles into Tuzla. The Globemaster III's flexibility was highlighted when the U.S. Army urgently needed additional pontoon bridging equipment in the area when the Sava River flooded

was subjected to a workout almost twice as strenuous. The C-17 participation in the airlift associated with Operation Joint Endeavor during the winter of 1995-96 marked the aircraft's first use in theater airlift missions and high-tempo emergency operations. The NATO effort brought peacekeepers into Bosnia-Herzegovina to separate the civil

war combatants and ensure a successful implementation of the Dayton peace treaty. The jet was also at work in the concurrent humanitarian relief supply airlift to Bosnia dubbed Operation Provide Promise. The first C-17 Provide Promise flight into Sarajevo was on 8 December. The 437th provided 12 C-17s (of only 19 in the inventory at the time) to a

only a quarter of the sorties. At its peak, a Globemaster III was landing at Tuzla or Taszar at a rate of one per hour in an around-the-clock operation. On 28 December alone, the C-17s carried 1.8 million pounds of cargo into Bosnia and Hungary. Over months of work in the Balkans the C-17s carried 44 percent of the cargo (19,892 tons) and 30 percent of the passengers (5,574) while flying only a quarter (26 percent) of the sorties (1,000). A C-17 also had the honor of being temporarily designated Air Force One when it ferried President Clinton on his visit to the American troops in the Balkans.

A C-17 pilot gently nudges his jet toward the contact position behind a KC-135. The Aerial Refueling (AR) receptacle would already have been opened, ready to be "plugged" by the tanker's boom. AR is one of the tasks requiring special training and qualification for C-17 pilots. (DoD)

and disrupted troop movement. Although never before loaded into the C-17, three of the jets carried 25 of the 32,000-pound sections. They were rolled aboard on flatbed trucks and flown to Taszar, in southern Hungary, where they were taken immediately to the Sava.

During the 60 days of Operation Joint Endeavor the C-17s flew 1,025 sorties and experienced a mission capable rate of 86.2 percent and a dispatch reliability rate of 97.8 percent (83.9 percent if weather delays are counted). This was accomplished while thousands of miles from support facilities and with no experience using war reserve maintenance kits. However, the C-17s required more than routine servicing between consecutive missions only two percent of the time, while the C-141s and C-5s averaged about 40 percent "break rate." The few problems that did crop up owing to the immaturity of the aircraft's systems were quickly dealt with and did not hamper operations.

Although the 17th Airlift Squadron's big jets performed only 18 percent of the relief missions into Bosnia they carried 30 percent of the cargo. For the entire operation the C-17s delivered more than all the other transports of Charlie Squadron combined (10 C-141s and 3 C-5s) and yet flew

The C-17 has done its share of emergency airlift in response to natural disasters. The first such instance was flying emergency relief supplies to the Virgin Islands following Hurricane Marilyn in mid-September 1995. The C-17 was the only heavy airlifter that could operate on the small surface area of the St. Thomas airport and the tremendous payload capacity of the jet made it more desirable than C-130s for the missions. The advanced navigation systems of the

Aircraft P-37 touches down with a puff of smoke from the tires. Note the bright metal on the slats inboard of the engines. This material better resists the heat from the core thrust reverser jet impingement. (USAF)

A C-17 copilot runs through preflight checks in preparation for a demonstration at the 1998 Farnborough Airshow. The officer is consulting one of the mission computer displays. Note the clear plastic wheel of the gear handle just above on the instrument panel. (DoD)

bases. In a grim operation, a C-17 flew to the area near Dubrovnik, Croatia, where Commerce Secretary Ron Brown and his party had been killed in a USAF T-43 crash in April 1996. Globemasters flew out the human remains as well as the entire aft fuselage of the wreckage. The jets returned to Africa in August 1998 with the solemn duty of bringing home the Americans killed in the embassy bombings.

Late November 1996 saw Globemaster IIIs in another trouble spot, this time central Africa. The aircraft flew to Entebbe, Uganda, with relief supplies as part of Operation Guardian Assistance. In March they were at Libreville, Gabon, with a team planning for the possible evacuation of Americans from the troubled Zaire.

The 17th Airlift Squadron's outstanding performance was recognized with the award of the General Joseph Smith trophy and by being named Air Mobility Command Airlift Squadron of the Year for 1996 and 1997.

In December 1995 two of the 437th AW's aircraft flew 147 U.S. Army Rangers from Georgia direct to Egypt where the soldiers parachuted to the desert below in support of the Bright Star '95 exercise. In another deployment exercise for CENTRAZ-BAT '97 the C-17 completed the longest-distance airborne mission ever. On 14 September 1997 eight aircraft departed Pope AFB, North Carolina, with 500 members of the 82nd Airborne Division for a nearly 20-hour, 8,000-nautical mile, nonstop flight. The soldiers parachuted onto a plateau near Shymkent, Kazakhstan. The jets returned to the former Soviet Union beginning on 28 October to help fly out disassembled MiG-29 fighters bought from Mol-

jet were also essential since all ground-based navigation aids on the islands were inoperative. The Globemaster III executed only 6 of the 50 sorties per day the U.S. Air Force flew into the islands yet delivered 20 to 30 percent of the cargo. In November 1998 the C-17s were again performing disaster relief services, this time in Central America following the Hurricane Mitch. The jets were busy again at the end of February 2000 bringing emergency supplies to flood-ravaged Mozambique.

Soon after the Virgin Islands emergency, the C-17s were called on to help support the evacuation of Americans and Europeans during the revolution in Liberia. The Globemasters flew giant CH-53 helicopters into Freetown, Sierra Leone, which in turn began immediately to fly evacuation missions. The C-17s also brought in supplies and support equipment while taking the evacuees out of the region. At the end of the operation the jets took all of the gear back to its European

davia so they could not be purchased by other parties.

An evacuation flight of another kind was flown on 9–10 September 1998 when P-38 flew Keiko, star of the *Free Willy* films, to Vestmannaeyjay Airport, Westman Islands, Iceland, from Newport, Oregon. Keiko is an Orca, weighing 10,000 pounds, who brought along over 70,000 pounds of "luggage" including its water-filled tank. Only the C-17 could fly such a load into the 3,900-foot runway on the remote island in Operation Keiko Lift.

When the Balkans heated up again in 1999 the much-enlarged Globemaster III fleet was again a primary airlifter in the 24 March to 20 June 1999 Operation Allied Force, NATO's campaign against Yugoslavia in reaction to its ethnic cleansing in Kosovo. Although still making up just a small percentage of the AMC's fleet, the C-17s flew 75 percent of the command's missions in the theater. During the several weeks of the campaign nearly all 50 machines of the existing C-17 fleet were employed in airlift missions, one aircraft sent overseas just two days after being delivered to the operational squadron.

The aircraft's capability to carry tremendous payload capacities into small, austere airfields and maneuver in the confined apron space proved exceedingly useful given the poor airfield facilities in Albania and Macedonia, and later Kosovo. In addition to the usual personnel and palletized cargo deliveries, the deployment of Task Force Hawk from Germany saw the C-17s airlifting tracked vehicles like the M1 tank, Multiple Launch Rocket System and Bradleys, other heavy weapons, and 2,000 troops. Apart

from the 24 AH-64 gunships, they also flew in Chinook and Black Hawk helicopters. At the peak, up to 22 C-17s were being flown into Albania each day and 468 Globemaster III sorties contributed to bringing in the entire force. This contingent was vital in the event the war expanded beyond the air campaign and its rapid deployment was due to the C-17's ability to operate at Albania's poorly serviced Tirana airfield.

Coincident with the offensive operation was Operation Sustain Hope, the airlift of humanitarian aid to the displaced Kosovar ethnic Albanians. The C-17 was instrumental in this quick-reaction airlift as part of Joint Task Force Shining Hope. The first C-17 flight into the region was on 4 April carrying 30,000 rations (72,000 pounds) in what grew to half a million daily rations and 700 tents. After the war, Operation Allied Force became Operation Joint Guardian with the C-17s remaining

in the theater and resupplying the U.S. contribution to the Kosovo Forces peacekeepers with 253 sorties, principally via Macedonia. By 29 June the Charleston AFB birds operating in the Balkans demonstrated a dispatch reliability of 96 percent for their 1,092 missions into the theater, much greater than the older airlifters. Later removal of Task Force Hawk and other U.S. elements in theater was an equally demanding airlift job.

The C-17 has become the DoD's airlifter of choice. Usually AMC customers simply request airlift support, but now they are stating that they want the C-17. The Globemaster III aircrews have given their jet the affectionate nickname of "Moose," reflecting its appearance from the front and the sound made by the wing vent boxes during ground refueling as air is forced out of the tanks. The name is also used as a call sign.

The pilot and scanner peer out intently as the C-17 approaches the Albanian coast, looking for any possible threats. Taken on 29 April 1999 during Operation Shining Hope, the jet is airlifting relief supplies to refugees from the war in Kosovo. The observer's seat may also be slid on tracks to the center of the cockpit for an evaluator to monitor crew performance. (DoD)

FUTURE POSSIBILITIES

Boeing sees a need for many more U.S. Air Force aircraft beyond those on contract in early 2000 and the Balkans experience appears to support this conclusion. In mid-1999, the company offered a guaranteed price of $149 million per copy for a further 60 C-17s. Congress authorized the U.S. Air Force to negotiate such a buy with the proviso that the "fly-away" price be 25 percent less than the average price of P-41 through P-120. The offer was still being negotiated at time of writing, but budget pressures may prevent future procurement.

Boeing has studied possible revisions to the C-17 to make it suitable for other missions, such as an airborne command post. The company has examined the advantages of stretching the airframe with a 12- to 40-foot fuselage plug. A stretch of 20 feet would still permit STOL operations, with more powerful engines allowing similar performance even with a longer plug. The stretch would permit an increase to 22 cargo pallets from 18, but would also impact the advantage of ground maneuverability in small ramp space and reduce freight throughput in some scenarios. The U.S. Air Force has shown no enthusiasm for stretching the C-17, but the option may come to the fore again should the service seek a C-5 replacement in the future.

The potential of the C-17 serving as the basis for the next generation U.S. Air Force tanker aircraft — a KC-17 — has been repeatedly mentioned. The intent to convert an existing airframe, military or commercial, to the tanker mission is clear and the C-17 would offer greater utility in the secondary airlift mission than the KC-135 or KC-10. The proposed configuration has a palletized Remote Aeri-al-Refueling Operator's station at the forward end of the cargo compartment. A flying boom would be installed in the cargo door along with an aerial refueling hose reel pod under each wingtip using a hardpoint already incorporated into the design. Cameras would be installed to permit the operator to monitor the refueling. A 6,740-gallon roll-on fuel tank could be secured in the cargo compartment to expand the C-17's already prodigious fuel capacity.

Boeing is seeking foreign orders for the C-17 to try to fill requirements for 24 aircraft. One stumbling block is the comparatively high unit cost of the aircraft of approximately $200 million. The C-17 competed with the Antonov An-124, Airbus A300-600F cargo conversion and A300-600ST Beluga to fill an RAF Short-Term Strategic Airlift requirement. The British initially found none of the contenders suitable or failing to satis-fy budget goals, but on 16 May 2000 decided to lease four Globemaster IIIs. Any further European sales will be competing with the notional Airbus A400M turboprop airlifter and the Antonov An-70, the U.K. already pledging support to the former. Australia and Japan are also seen as potential markets for the C-17. The possibility of fitting the Rolls-Royce RB211 engine to the C-17 has been mentioned since early in the program and may yet make the aircraft more attractive to some customers.

Boeing has also continuously emphasized the potential of the C-17 for commercial cargo operations and has explored the possibility of leasing a U.S. Air Force C-17 to demonstrate the practicality of such operations. Although Boeing has projected a need for more than 1,200 new freighters through 2017, the proposed MD-17™ has found no customers to date.

A U.S. Air Force security policeman provides airfield security at Tirana on 11 April 1999 during Operation Sustain Hope. Beyond, personnel are rolling pallets of food, water, and medicines from the Globemaster III onto a K-loader while, due to the urgency, a forklift extracts the loads from the opposite end. (DoD)

C-17A PRODUCTION LIST

ACTUAL AND PROGRAMMED AS OF 2000

A/C	Serial Number	Delivery Date	Notes
T-1	87-0025	15 Sep 91	dedicated to test
	Lot I (2 aircraft)		
P-1	88-0265	18 May 92	initially a flight test asset
P-2	88-0266	21 Jun 92	initially a flight test asset
	Lot II (4 aircraft)		
P-3	89-1189	7 Sep 92	initially a flight test asset
P-4	89-1190	9 Dec 92	initially a flight test asset
P-5	89-1191	12 Mar 93	initially a flight test asset
P-6	89-1192	10 Jun 93	initial operational a/c, *The Spirit of Charleston*
	Lot III (4 aircraft)		
P-7	90-0532	26 Aug 93	first with Airlift Defensive System
P-8	90-0533	29 Oct 93	
P-9	90-0534	30 Dec 93	
P-10	90-0535	8 Feb 94	
	Lot IV (5 aircraft)		
P-11	92-3291	8 Apr 94	
P-12	92-3292	18 May 94	
P-13	92-3293	29 Jun 94	
P-14	92-3294	20 Aug 94	
P-15	93-0599	29 Sep 94	
	Lot V (5 aircraft)		
P-16	93-0600	18 Nov 94	
P-17	93-0601	22 Dec 94	
P-18	93-0602	17 Feb 95	*City of Altus*
P-19	93-0603	14 Apr 95	
P-20	93-0604	19 Jun 95	first with 94 Package engine improvements
	Lot VI (6 aircraft)		
P-21	94-0065	31 Jul 95	
P-22	94-0066	29 Sep 95	
P-23	94-0067	21 Nov 95	
P-24	94-0068	21 Jan 96	
P-25	94-0069	1 April 96	
P-26	94-0070	31 May 96	
	Lot VII (6 aircraft)		Block 6 upgrades
P-27	95-0102	3 Jul 96	*Spirit of Long Beach*
P-28	95-0103	29 Aug 96	
P-29	95-0104	15 Nov 96	
P-30	95-0105	14 Jan 97	first with cruise performance enhancements
P-31	95-0106	25 Mar 97	*Spirit of Bob Hope*
P-32	95-0107	23 May 97	*Spirit of North Charleston*
	Lot VIII (8 aircraft)		Block 8 upgrades
P-33	96-0001	28 Aug 97	first with new gear pods
P-34	96-0002	30 Sep 97	*Spirit of The Air Force*
P-35	96-0003	13 Nov 97	
P-36	96-0004	23 Dec 97	
P-37	96-0005	29 Jan 98	*Spirit of Sgt. Levitow*

A/C	Serial Number	Delivery Date	Notes
P-38	96-0006	28 Feb 98	*Spirit of Berlin*
P-39	96-0007	11 Apr 98	*Spirit of America's Veterans*
P-40	96-0008	15 May 98	*Spirit of Total Force*
	Lot IX (8 aircraft)		Block 9 upgrades S/Ns aligned with production numbers
P-41	97-0041	10 Aug 98	first with N/EAT nacelles & new litter stanchions
P-42	97-0042	18 Sep 98	*Spirit of The Tuskegee Airmen*
P-43	97-0043	17 Oct 98	*Spirit of Los Angeles*
P-44	97-0044	7 Nov 98	
P-45	97-0045	12 Dec 98	
P-46	97-0046	22 Dec 98	*The Spirit of The Military Family*
P-47	97-0047	18 Feb 99	*Spirit of Middle Georgia*
P-48	97-0048	20 Mar 99	
	Lot X (9 aircraft)		Block 10 upgrades
P-49	98-0049	22 Apr 99	
P-50	98-0050	18 May 99	
P-51	98-0051	25 Jun 99	first with composite tail
P-52	98-0052	26 Jul 99	*Spirit of McChord*
P-53	98-0053	25 Aug 99	
P-54	98-0054	17 Sep 99	
P-55	98-0055	22 Oct 99	
P-56	98-0056	4 Dec 99	
P-57	98-0057	11 Dec 99	
	Lot XI (13 aircraft)		Block 11 upgrades
P-58	99-0058	17 Mar 00	first with dual-row airdrop capability
P-59	99-0059	31 Mar 00	
P-60	99-0060	31 Mar 00	
P-61	99-0061	18 May 00	
P-62	99-0062	15 Jun 00	Spirit of The Hump
P-63	99-0063	30 Jun 00	Spirit of the Wright Brothers
P-64	99-0064	3 Aug 00	The City of Saint Louis
P-65	99-0065	31 Aug 00	
P-66	99-0066	11 Sep 00	
P-67	99-0067	22 Sep 00	
P-68	99-0068	2000	
P-69	00-0069	2000	
P-70	00-0070	2000	
	Lot XII (15 aircraft, 3 aircraft later deferred to 2003)		
P-71	00-0071	2001	first with extended range capability
P-72	00-0072	2001	
P-73	00-0073	2001	
P-74	00-0074	2001	
P-75	00-0075	2001	
P-76	00-0076	2001	
P-77	00-0077	2001	
P-78	00-0078	2001	
P-79	00-0079	2001	
P-80	00-0080	2001	
P-81	00-0081	2001	
P-82	00-0082	2001	
P-83	00-0083	2001	
P-84	00-0084	2001	
P-85	00-0085	2001	

C-17A PRODUCTION LIST

A/C	Serial Number	Delivery Date	Notes
Lot XIII (15 aircraft)			
P-86	01-0086	2002	
P-87	01-0087	2002	
P-88	01-0088	2002	
P-89	01-0089	2002	
P-90	01-0090	2002	
P-91	01-0091	2002	
P-92	01-0092	2002	
P-93	01-0093	2002	
P-94	01-0094	2002	
P-95	01-0095	2002	
P-96	01-0096	2002	
P-97	01-0097	2002	
P-98	01-0098	2002	
P-99	01-0099	2002	
P-100	01-0100	2002	
Lot XIV (15 aircraft)			
P-101	02-0101	2003	
P-102	02-0102	2003	
P-103	02-0103	2003	
P-104	02-0104	2003	
P-105	02-0105	2003	
P-106	02-0106	2003	
P-107	02-0107	2003	
P-108	02-0108	2003	
P-109	02-0109	2003	
P-110	02-0110	2003	
P-111	02-0111	2003	
P-112	02-0112	2003	
P-113	02-0113	2003	
P-114	02-0114	2003	
P-115	02-0115	2003	

A/C	Serial Number	Delivery Date	Notes
Lot XV (5 aircraft)			
P-116	03-0116	2004	
P-117	03-0117	2004	
P-118	03-0118	2004	
P-119	03-0119	2004	
P-120	03-0120	2004	
Lot XVI (6 aircraft, not funded at time of writing)			
P-121	04-0121	2005	
P-122	04-0122	2005	
P-123	04-0123	2005	
P-124	04-0124	2005	
P-125	04-0125	2005	
P-126	04-0126	2005	
Lot XVII (9 aircraft, not funded at time of writing)			
P-127	05-0127	2006	
P-128	05-0128	2006	
P-129	05-0129	2006	
P-130	05-0130	2006	
P-131	05-0131	2006	
P-132	05-0132	2006	
P-133	05-0133	2006	
P-134	05-0134	2006	
P-135	05-0135	2006	or possibly 2007 delivery

(tail number consists of the last five digits of serial number)

From this perspective it is clear how the C-17 earned its nickname "Moose." Another reason is the noise made by air escaping from the wing tank vent boxes during ground refueling. The jet may look ungainly at times but has proven itself an exceptionally capable airlifter and a vital addition to the U.S. Air Force's fleet. (AFFTC)

C-17A SPECIFICATIONS

Dimensions

Length: 174 ft. (53.04 m.)

Fuselage diameter: 22.5 ft. (6.86 m.)

Aft fuselage up-sweep: 15°

Height to top of vertical stabilizer: 55.1 ft. (16.79 m.)

Height at top of forward fuselage: 23 ft. (7.01 m.)

Wingspan: 169.8 ft. (51.74 m.)

Wing area: 3,800 ft.^{2} (353.03 m.^{2})

Wing aspect ratio: 7.165

Wing quarter-chord sweep angle: 25°

Wing anhedral angle: 3°

Wingtip height off of the ground (nominal): 13.8 ft. (4.2 m.)

Winglet angles: 30° aft sweep, 15° outboard

Winglet span: 9.21 ft. (2.81 m.)

Winglet height: 8.92 ft. (2.72 m.)

Winglet area: 35.85 ft.^{2} (3.33 m.^{2})

Winglet aspect ratio: 2.367

Horizontal stabilizer span: 65 ft. (19.81 m.)

Horizontal stabilizer area: 845 ft.^{2} (78.50 m.^{2})

Horizontal stabilizer aspect ratio: 5.0

Horizontal stabilizer leading edge sweep angle: 27°

Horizontal stabilizer anhedral angle: 3°

Vertical stabilizer area: 685 ft.^{2} (63.64 m.^{2})

Vertical stabilizer sweep angle: 41°

Wheelbase: 65.8 ft. (20.06 m.)

Wheel track: 33.67 ft. (10.26 m.)

Tire diameter: main 50 in., nose 40 in.

Nose gear steering angles: ±12° via rudder pedals, ±65° via tiller

Engine fan diameter: 85 in.

Control Surface Angles

Aileron rotation angles: 40° up, 27° down

Spoiler rotation angles: 11° to 16° as speedbrakes, 60° as spoilers

Slat segment extension angles: 18° inboard, 23° center,
25° outboard pair

Horizontal stabilizer trim authority angles: 12° up, 4° down

Elevator rotation angles: 25° up, 15° down

Rudder rotation angles:
forward segment ±28.2° (±25° with EFCS off)
aft segment ±26° (relative to the forward segment)

Flap deflection angle: 40.5° full

Flap deflection rate: 2 deg./sec. extend, 3 deg./sec. retract

Cargo Loading Specifics

Cargo compartment floor height off ground: 5.33 ft. (1.62 m.)

Cargo ramp angle to ground: 9°

Cargo ramp length: 19.8 ft. (6.04 m.)

Cargo compartment floor length: 68.2 ft. (20.78 m.)

Cargo compartment loadable length
(including ramp): 85.17 ft. (25.96 m.)

Cargo compartment loadable width: 18 ft. (5.49 m.)

Cargo compartment loadable height: 13.42 ft. (4.09 m.)
(under wing) 12.33 ft. (3.76 m.)

Cargo compartment volume: 20,900 ft.^{3} (591.47 m.^{3})

Performance

Crew accommodations: 2 pilots, 1 loadmaster,
2 additional crewmembers (optional)

Passenger accommodations: 102 troop seats, 36 litters

Powerplant: four 40,400 lb.-thrust P&W F117-PW-100 turbofans

Maximum grade for backing: 2%

Minimum 180° ground turn radius: 116 ft. (35.4 m.), 85 ft. (25.9 m.)
using "star turn" tighter when towed

Landing gear cycle time: 10 sec. extend, 13 sec. retract

Maximum usable fuel load (JP-8): 181,054 lb. (82,125 kg.)

Maximum cargo: 169,000 lb. (76,657 kg.) 170,900 lb. (77,519 kg.)

Maximum cargo to 4,000 Nmi (unrefueled): 57,000 lb. (25,854 kg.)

Ramp load capacity: 40,000 lb. (18,144 kg.)

Operating (empty) weight: 277,000 lb. (125,645 kg.)

Maximum ramp weight: 586,000 lb. (265,805 kg.)

Maximum takeoff/landing weight: 585,000 lb. (265,351 kg.)

Cruise speed: 0.74 to 0.77 Mach

Maximum speed (V_H): 350 knots, 0.825 Mach

Limit speed (V_L): 398 knots, 0.875 Mach

Maximum g: +3.0 and -1.0 at 418,000 lb. (189,602 kg.)

Service ceiling: 45,000 ft. (13,716 m.)

Takeoff field length (max. gross weight): 7,740 ft. (2,359.15 m.)

Landing field length: (160,000 lb. payload): 3,000 ft. (823 m.)

Unrefueled range:
(160,000 lb. payload): 2,400 Nmi (4,442 km.)
(130,000 lb. payload): 3,200 Nmi (5,926 km.)
(40,000 lb. payload - paratroops): 4,400 Nmi (8,149 km.)
(empty – ferry): 4,700 Nmi (8,704 km.)

Significant Dates

15 October 1980
C-X request for proposal issued

28 August 1981
McDonnell Douglas Model D-9000 selected C-X winner

23 July 1982
Low-level development contract awarded

31 December 1985
C-17 Full Scale Engineering Development contract let

2 November 1987
First C-17 part manufactured

24 October 1988
First C-17 parts assembled

15 September 1991
First flight and delivery of a C-17A (aircraft T-1)

3 October 1991
First aircraft backing operation (T-1)

17 October 1991
First flight and landing in full mechanical reversion (T-1); First low speed flight at full flaps (T-1)

22 November 1991
First engine airstart (flight test, T-1)

17 January 1992
First 100 hours of flight time logged (T-1)

11 April 1992
First aerial refueling (KC-135 and T-1)

22 April 1992
First flight above 40,000 feet (flight test, T-1)

29 May 1992
First flight to limit Mach number (flight test, T-1)

3 June 1992
First flight to limit calibrated airspeed (flight test)

10 June 1992
First in-flight thrust reverse (flight test, T-1)

17 June 1992
First in-flight opening of cargo door and ramp (flight test, T-1)

13 August 1992
First in-flight opening of paratroop doors (flight test, T-1)

29 August 1992
100th flight of a C-17 (T-1)

September 1992
First onload of tracked vehicle (M-60 tank)

7 September 1992
First landing on unpaved surface (P-3)

15 September 1992
First formation flight (2-, 3-, and 4-ship)

23 September 1992
First aerial refueling from a KC-10 (flight test, T-1); First night flight, takeoff and landing (flight test, P-2)

4 February 1993
First 1,000 hours of flight time logged; First airdrop – door bundle (flight test, P-1)

3 May 1993
First LVAD airdrop (flight test, P-1)

14 June 1993
First delivery of a C-17 (P-6) to an operational unit (17th AS)

9 July 1993
First paratroop jump – HALO (flight test, T-1)

23 August 1993
First night HALO jump (flight test, P-4)

September 1993
2,000th flight hour logged on C-17 fleet

10 November 1993
First night aerial refueling (flight test, P-5)

11 November 1993
First paratroop jumps (static line) out troop doors (flight test, P-1)

9 December 1993
First steep approach landing (flight test, P-2); First double-stick CDS airdrop (flight test, P-1)

8 April 1994
First LVAD sequential platform airdrop (flight test, T-1)

3 May 1994
First LAPES delivery (flight test, P-1)

September 1994
First airdrop of a 60,000-pound platform

October 1994
First maximum 18 463L pallets combat offload

14 October 1994
First operational mission, Operation Vigilant Warrior (supporting airlift to Kuwait)

16 December 1994
Development test and evaluation completed

17 January 1995
First squadron (17th AS) declared operational (12 aircraft)

28 April 1995
First 102-man paratroop drop (flight test)

27 June 1995
First formation airdrop (6 aircraft), 55 tons (including Sheridan light tank) and 204 paratroopers

5 August 1995
30-day Reliability, Maintainability, and Availability evaluation completed

3 November 1995
Decision made to execute full 120-aircraft buy of C-17s

31 May 1996
Multi-year contract for 80 C-17s signed

9 April 1997
First large-formation airdrop exercise (9 C-17s and 9 C-141s)

6 September 1997
First aeromedical evacuation flight – new litter stanchions (exercise)

Summer 1998
First 100,000 hours of flight time logged

18 May 1999
Delivery of 50th production aircraft

16 May 2000
United Kingdom announces decision to lease four C-17s